T0229928

THE AUTONOMY OF
MATHEMATICAL KNOWLEDGE

Most scholars think of David Hilbert's program as the most demanding and ideologically motivated attempt to provide a foundation for mathematics, and, because they see technical obstacles in the way of realizing the program's goals, they regard it as a failure. Against this view, Curtis Franks argues that Hilbert's deepest and most central insight was that mathematical techniques and practices do not need grounding in any philosophical principles. He weaves together an original historical account, philosophical analysis, and his own development of the meta-mathematics of weak systems of arithmetic to show that the true philosophical significance of Hilbert's program is that it makes the autonomy of mathematics evident. The result is a vision of the early history of modern logic that highlights the rich interaction between its conceptual problems and its technical development.

CURTIS FRANKS is Assistant Professor in the Department of Philosophy, The University of Notre Dame.

THE AUTONOMY OF MATHEMATICAL KNOWLEDGE

Hilbert's Program Revisited

CURTIS FRANKS

University of Notre Dame

CAMBRIDGE
UNIVERSITY PRESS

CAMBRIDGE UNIVERSITY PRESS
Cambridge, New York, Melbourne, Madrid, Cape Town,
Singapore, São Paulo, Delhi, Tokyo, Mexico City

Cambridge University Press
The Edinburgh Building, Cambridge CB2 8RU, UK

Published in the United States of America by Cambridge University Press, New York

www.cambridge.org
Information on this title: www.cambridge.org/9780521514378

First published 2009
First paperback edition 2010

A catalogue record for this publication is available from the British Library

ISBN 978-0-521-51437-8 Hardback
ISBN 978-0-521-18389-5 Paperback

For Lydia

The one and what I said about it make two,
and two and the original one make three.
If we go on in this way,
then even the cleverest mathematician can't tell
    where we'll end,
much less an ordinary man.

If by moving from nonbeing to being we get to three,
how far will we get if we move from being to being?
Better not to move, but to let things be!

Chuang Tzu, *The Inner Chapters*

Contents

Preface

When I decided to bundle my recent historical, philosophical, and logical research together into a book, I considered two different approaches. One approach was to try to represent David Hilbert's foundational program exhaustively, to let my own findings simply shape the re-telling of a fairly familiar story. Another approach — the one that I ultimately preferred — was to center the book around what I think are both the most important and most overlooked aspects of Hilbert's program. There is no lack of high quality writing about Hilbert, nor of high quality development of his scientific innovations, but there seems to me to be a glaring oversight of one truly unique aspect of Hilbert's thought. So this book is a modest attempt to isolate, explain, and develop a single strain of Hilbert's philosophy. If the book inspires new interest and appreciation of Hilbert, then it has served its purpose.

The core of the book grew out of my doctoral research at the University of California's Department of Logic and Philosophy of Science in Irvine. The title of my doctoral thesis, "Mathematics speaks for itself," was a *double entendre*. It was meant to suggest two distinct but related themes. The first theme is that questions about mathematics that arise in philosophical reflection — questions about how and why its methods work — might be best addressed mathematically. I believe that this is so, and I claim that David Hilbert held the same view. In Chapter 2, I explain that Hilbert's program was primarily an effort to demonstrate that mathematics could answer questions about how its own methods work. Hilbert thought that if he could succeed at this, then he would have carved out a privileged position for mathematics among the sciences. Unlike other ways of knowing, the validity of ordinary mathematical methods would be seen to be independent of any philosophical theories of knowledge, as autonomous.

The second theme arises out of the first. Once one sees mathematics potentially providing its own foundations, one faces questions about the available ways for it to do so. The two most poignant issues are how a formal theory should refer to itself and how properties *about* a theory should be represented within that theory. I claim that standard answers to these questions are not sufficiently free from extra-mathematical, philosophical assumptions to speak fully to Hilbert's vision. In Chapter 3, I examine Hilbert's own attempt to settle these questions and explain why his attempt failed by his own standards. Then I turn to Jacques Herbrand's contribution to Hilbert's program and discuss the partial progress he made to Hilbert's goal of mathematical autonomy. Herbrand's work is known primarily for its purely mathematical accomplishments. I do not know of any detailed study on his philosophical perspective. However, a close look at his methods and remarks about the significance of his results reveals that he had a rich philosophical perspective, close in spirit to Hilbert's. In fact he viewed his own [1930a] Fundamental Theorem as a contribution to Hilbert's project of formulating questions of metatheory purely mathematically, and he even recognized the need for additional techniques of arithmetization (of the sort Gödel would later supply) to complement his own.

In Chapter 4, I rephrase the discussion in terms due to Solomon Feferman. It turns out that his notion of *intensionality* is precisely what a mathematical study of mathematics – in the spirit of Hilbert's original vision – requires. I examine the extent to which Gödel's and Herbrand's techniques of arithmetization are intensionally correct and suggest that a certain combination of the two works much better than either one on its own. Specifically, a Gödel-style encoding of the formulations of provability and consistency that result from Herbrand's theorem returns formulas that can be relativized to the computational strength of any arithmetical system. As a result one is able to pose questions about a system's metatheory to that system always in a way such that the system can understand the questions.

As an illustration of the applicability of these techniques, in Chapter 5 I apply the point of view from previous chapters to a specific problem in the philosophy of mathematics: whether the fact that a version of Gödel's second incompleteness theorem for Robinson's arithmetical theory Q can be understood as showing that Q cannot prove its own consistency. It is worth mentioning the chapter's focus on weak mathematical systems. I have often heard philosophers bemoan the attention that mathematical logic research, especially in recent years, gives to weak theories. My focus is on bounded and induction-free fragments of arithmetic, which are weak systems by any standard. Thus I must explain myself. I am in full sympathy with philosophers who see foundational studies as missing an essential point when focused solely on such weak systems. Weak fragments of arithmetic are not the theories that mathematicians ordinarily use, so it seems at first odd to suggest that a study of these theories can turn up "a foundation of mathematics." A singular focus on weak arithmetics appears to most philosophers just as the obscure preoccupations of his day appeared to Laurence Sterne, as "common-place infirmity of the greatest mathematicians! working with might and main at the demonstration, and so wasting all their strength upon it, that they have none left in them to draw the corollary, to do good with" (Sterne [1759–67], pp. 87–8).

But I propose to draw that corollary. In the spirit of Hilbert's program, my project is not to provide epistemological foundations for the strongest system I can, starting from the bottom up, only to stop there and advocate a restriction of mathematical methods to those so founded. Rather, with Hilbert, I am interested in a system's ability to refer to itself and thereby to demonstrate properties that it has. Since a system's ability to perform these tasks depends on its strength, a precise study of the phenomenon involves studying systems of different strengths. Weak arithmetic theories admit different arithmetization schemes and therefore perform these tasks in different ways. Consequently they are the natural place to turn to investigate how mathematics speaks for itself.

As I said, the strain of Hilbert's thought that I isolate and develop in this book is not ordinarily associated with Hilbert's program. Nevertheless, I believe that once it is recovered, it complements many other, well-known features of Hilbert's thought. But more glaringly, I believe that it runs directly counter to the ideological position that is often, but erroneously, attributed to Hilbert. In Chapters 1 and 6 I focus on the significance of Hilbert's anti-foundational stance, and try to rethink how his philosophical ideas fit with the historical events that provided their context as well as how they have influenced contemporary philosophy of mathematics.

Acknowledgments

For such a short book as this, I have a long list of debts.

A great deal of the content of this book was inspired by conversations with colleagues during the last half decade or so. Most obviously, my doctoral committee at the University of California's Department of Logic and Philosophy of Science in Irvine not only educated me in the relevant philosophical history and formal techniques, but helped shaped my own views as I was formulating them. This committee consisted of Aldo Antonelli, Jeffrey Barrett, and Kai Wehmeier. I learned a great deal from them all about how to approach philosophical problems. Aldo's guidance in particular, at every stage, was essential. It is not possible to reconstruct the several ways that his advice has shaped this project.

My debt extends well beyond this circle, though, to other specialists who contributed to my understanding of the various philosophical, mathematical, historical, and linguistic topics that arise in this book. Advice from Michael Detlefsen, Penelope Maddy, Volker Halbach, Neil Delany, Richard Grandy, Sam Buss, Richard Rorty, and Thomas Saine has been particularly helpful. Again, the scarcity of specific acknowledgments to these colleagues should be understood as evidence of the depth of their influence.

I owe an equally great debt to those who most encouraged me to turn this project into a book. I share the credit for its publication with Aldo Antonelli, Jeffrey Barrett, Michael Detlefsen, Patricia Blanchette, Richard Rorty, Leo Corry, Andrew Boucher, Hilary Gaskin, and two anonymous reviewers of an earlier manuscript. Without their vision, the book not only would have not taken on the shape that it has taken, but would not likely have appeared in any form.

I am deeply grateful to my students at the University of Notre Dame, who have often been the philosophical audience that pruned and sharpened what I wanted to say. Conversations with Graham Leach-Krouse, Charles

Pence, John Firth, Sean Walsh, Tony Strimple, Jon Buttaci, Louis Gularte, and Alana Stelton were especially valuable.

Among friends I have mentioned twice, both for their contributions to this project and for their vision of its potential, is Richard Rorty, whose advice and optimism, though occasional, were essential. I regret being unable to share this book with him.

I presented an earlier version of Chapter 2 at the *GAP* conference in Berlin in 2006. I regret that for unfortunate legal reasons I was not able to accept Benedict Lowe's invitation to contribute the chapter to the proceedings of his wonderful session, "Towards a new epistemology of mathematics." An earlier version of Chapter 3 was the basis of my talk at the 2008 annual meeting of the Association of Symbolic Logic in Irvine. Chapter 4 grew out of a presentation that I made to a graduate seminar on Gödel that Michael Detlefsen and I organized in 2007, and improvements to Chapter 6 grew out of discussions in my undergraduate seminar "Philosophy against itself" in 2008. I presented an early version of Chapter 5 at the Notre Dame Mathematics Department's Logic colloquium in 2007. I thank the members of these various audiences and seminars, many of whom are not already mentioned above.

CHAPTER 1

A new science

1.1   RECOVERING HILBERT'S THOUGHT

No one disposed to judge the worth of an idea by its impact on culture through contributions to art and science would too quickly dismiss David Hilbert's philosophy of mathematics. Although he worked in an era when mathematicians were especially prone to reflect on the nature of their discipline, when philosophies of mathematics numbered nearly as many as great mathematical minds, the innovative research spawned by Hilbert's unique views stands out for its lasting imprint on mathematical practice. Yet oddly, few people today endorse his views. In the main, they are deplored.

I find this paradox intolerable, and I hope to dissolve it by unearthing its origins. This will be somewhat arduous, but it is worth the effort. Hilbert's ideas have not been rejected because of their faults, but because his true vision is unknown. The excavation that follows will, I hope, expose the genius of his philosophical vision and its essential connection to his mathematical innovations.

In the early twentieth century, Hilbert invented a new formal science – the study of the global properties of branches of mathematics like number theory, analysis, and group theory. This invention allowed one for the first time to investigate in a scientific manner whether, for example, the principles used by analysts are consistent or whether, to take another example, the principles used by group theorists suffice to answer all questions about groups. Hilbert's vision involved two steps. First he explained how to isolate the definitive principles of a branch of mathematics and design a formal system (a set of axioms and inference rules) that embodies those principles. Then he explained how reasoning mathematically about the combinatorial properties of these formal systems leads to the discovery of facts about the entire class of theorems that can be proved with them. Hilbert sometimes called this science "proof theory," other times "meta-mathematics."

A century later, meta-mathematics is a thriving discipline. In addition to the continued interest in questions that have arisen directly out of its development, meta-mathematics has shed insight into some of the deepest problems in computer science, logic, and main-line mathematical research. Hilbert's pioneer efforts in the field have thus proved to be a great achievement. If the significance and beauty of a science speaks for the validity of the ideas that it grew out of, then the views that spawned his efforts have been emphatically vindicated. They bear the rare mark of philosophical genius, the customs stamp signifying their safe landing on science's soil.

One would like to understand Hilbert's philosophical views, to gain some insight into the formative moments of an exciting modern science, even to align one's own understanding of mathematics with his. However, two attitudes dominate contemporary discussion of Hilbert's thought, and their influence screens from historical access the revolutionary insight that led him to forge his new science. The more dominant of these attitudes was probably most forcefully voiced by Alfred Tarski. As early as 1930 Tarski emphasized that the establishment of meta-mathematics as an independent branch of mathematics liberated it from its conceptual origins. Meta-mathematical concepts, he explained in [1931], "do not differ at all from other mathematical notions" so that "their study remains entirely within the domain of normal mathematical reasoning" (p. 111). Specifically he declared that Hilbert's alleged hope that meta-mathematics would usher in a "feeling of absolute security" was "a kind of theology" that lay "far beyond the reach of any normal human science" ([1995], p. 160). Since Hilbert's principal accomplishment, as Tarski saw it, was that his efforts had solidified into a normal human science, the conceptual framework that gave rise to them – being arcane and ideological – is both irrelevant to the further pursuit of meta-mathematics and erroneous as a portrayal of the field's nature. On the same grounds, G. H. Hardy began his famous caricature of Hilbert's thought in his 1928 Rouse Ball lecture with these splintering words:

I find it very necessary to distinguish between Hilbert the philosopher and Hilbert the mathematician. I dislike Hilbert's philosophy quite as much as I dislike that of Brouwer and Weyl, but I see no reason for supposing that the importance of his logic depends in any way on his philosophy. ([1929], p. 1251)

Opposing this Tarskian tenor is a second, more favorable attitude towards Hilbert's thought. Scholars with this attitude think that Hilbert's philosophical views are well worth taking seriously. They see Hilbert holding an irrealist or nominalist view about the nature of mathematics called "formalism," and espousing "finitism," the skeptical view that only a highly restricted class of mathematical techniques are *prima facie* legitimate. Although these scholars disagree about exactly how Hilbert's formalist and finitist views figure into the vision he had set for meta-mathematics,[1] they agree that he intended to provide an epistemological foundation for mathematics

---

[1] Two points of contention are the place of conservativity results in Hilbert's program and the status (as meaningful or merely instrumental constructions) of ideal elements. Raatikainen [2003], pp. 166–9 and Mancosu [1998a], pp. 159–61 survey these debates.

in a way that was informed by these views. Yet they tend also to agree in a certain sense with Tarski's idea that Hilbert's foundational aims have been separated from meta-mathematical research in the very process of that body of research attaining the status of an ordinary human science. Thus despite a deep affinity for foundational matters, Georg Kreisel explains that

the passage *from* the foundational aims for which various branches of modern logic were originally developed *to* the discovery of areas and problems for which its methods are effective tools ... did not consist of successive refinements, a gradual evolution by adaptation ..., but required radical changes of direction, to be compared to evolution by migration. ([1985], p. 139)

Kreisel's metaphorical language suggests that logicians had to shed their foundational aspirations in order to enjoy the full flowering of meta-mathematics as a science.

Clearly, neither attitude is of much help for understanding how meta-mathematics came to be. If the philosophical demands that Hilbert placed on his studies no longer shape meta-mathematics in its mature form, then one must turn elsewhere to find the field's true conceptual origins. But it is disingenuous to deny the tremendous formative impact of Hilbert's early proof-theoretical investigations, however entangled these may have been in a grander program. And worse, if meta-mathematics achieved the status of normal science in the very process of its practitioners shifting their attention away from philosophical goals, then there may be no instructive story of its invention to tell — its chief engineers having merely rescued some *accidental* features of an overly ambitious program by showing that these features could function miraculously on their own.

These scholarly attitudes towards Hilbert's philosophical thought have their own conceptual history, though. They could very well have not developed but for the entrenchment of a well-rehearsed story about how Hilbert's chief aims were dashed by two of the earliest meta-mathematical results, Kurt Gödel's incompleteness theorems. According to this story, Hilbert had designed his proof theory in order to demonstrate that the principles of classical mathematics were formally consistent, free from the threat of paradox. Moreover, since this demonstration was supposed to be carried out according to the constraints of formalism and finitism, the ensuing defense of classical mathematics would also serve as evidence that finitary reasoning about formal signs was the ultimate foundation of mathematical activity. Thus the "absolute certainty" that Tarski mocked is to be found in the security of this elementary form of reasoning. But since the incompleteness theorems demonstrate the unavailability of the kind of result Hilbert sought, Hilbert's philosophical views are not only beyond the reach of human science but also clearly erroneous. This story takes on a dramatic tone in the irony it depicts: The incompleteness theorems were among the earliest results to draw significant attention to the then fledgling discipline, so Hilbert's philosophical vision was toppled by the same blow that hammered his technical program into a permanent science.

Seen as reactions to this story, the current opinions about Hilbert's thought begin to make sense. Those sympathetic to formalism and finitism as philosophical doctrines happily let meta-mathematics continue on its course uninformed by such scruples, embracing an attitude satirized by Richard Rorty:

In every generation, brilliant and feckless philosophical naifs ... turn from their own specialties to expose the barrenness of academic philosophy and to explain how some or all of the old philosophical problems will yield to insights gained outside philosophy – only to have the philosophy professors wearily explain that nothing has changed at all. ([1976], p. 32)

Since meta-mathematics left its philosophical roots behind in its passage to normal science, those roots cannot be condemned by results, like Gödel's, issuing from meta-mathematics itself. Finitism and formalism live on, and it is up to philosophical deliberation, not the tribunal of mathematical theorems, to determine their ultimate merits or faults! Meanwhile the less sympathetic, Tarski among them, were more drawn to the elegance of the emerging science than they were concerned with the fate of its conceptual origins. If Hilbert's philosophical views had not been undermined by the verdict issued by his own invention – if they *cannot* be undermined because they are too philosophically pure to need to answer to something as mundane as a scientific result – then foundationalism is merely modern logic's quaint, embarrassing heritage. Hilbert invented meta-mathematics, but its enduring self-image derives from Tarski!

This situation poses an inevitable question: If Hilbert's philosophical ideas were so bad, how could they have been so scientifically productive? For all the compelling drama it depicts, the suggestion that a mathematical discipline should grow out of revolutionary insight, only for later development of the new mathematics to expose that insight's hopelessness and error, is surely implausible. Neither is it any help to try to salvage Hilbert's vision from the wreckage by spiriting it away into philosophically pure, closed-off quarters. That only compounds the mystery. Ideas that are scientifically inert by design ought to yield even less fruit than mistaken but application-oriented ones.

The fact that the attitudes canvassed above lead so naturally to this question explains the difficulty in understanding Hilbert's meta-mathematical revolution. But the dilemma this question poses is no reason to despair. The question is unanswerable because the "bad ideas" that one cannot envision hooking up with their mathematical fruits are distortions of Hilbert's actual thought. In stark contrast to the interpretation that attributes these ideas to him, I see Hilbert's genius stemming precisely from the fact that his ideological inclination was whole-heartedly scientific. His philosophical strength was not in his ability to carve out a position among others about the nature of mathematics, but in his realization that the mathematical techniques already in place suffice to answer questions *about* those techniques — questions that rival thinkers had assumed were the exclusive province of pure philosophy. The conceptual framework that Hilbert continuously referenced in his early proof-theoretical writings was not a hermetic landscape, to be evaluated by the sublimity of the arguments in its favor. Hilbert designed his research program to strip the fate of mathematics from the edict of those who would pronounce on its "true nature" and redeposit it in the hands of the scientific community. To understand the subtlety of his ideas and the way meta-mathematics emerged from them, one must count Hilbert among Rorty's "philosophical naifs." One must see him deliberately offering mathematical explanations where philosophical ones were wanted. He did this, not to provide philosophical foundations, but to liberate mathematics from any apparent need for them. "Defending" Hilbert's ideas by claiming that they are untouched by scientific findings insults that vision. It is, to borrow another of Rorty's quips, "like complimenting a judge on his wise decision by leaving him a fat tip" or like trying to praise a postmodernist by telling him that his views exhibit all of modernity's signs of truth ([1979], p. 372). The legitimacy of Hilbert's philosophical stance lies precisely in its ability to generate an arena for the scientific study of mathematics. Thus the above question inverts. Since Hilbert's philosophical ideas have been so scientifically productive, they must have been quite good. What were they?

To answer *this* question, one must shed the image of Hilbert as a dogmatic finitist and formalist. Then a fresh look at Hilbert's views, as he and as his philosophical colleague and spokesman Paul Bernays explain them in a series of essays and scientific reports in the 1920s, reveals three things. First, it shows how different Hilbert's thinking was from that of his ideological adversaries. It also shows that his thinking was remarkably different from the image that, as a result of a century of rhetoric in line with Tarski's statements, has become a fixture in discussions of the philosophy of mathematics – the image of Hilbert trying to ground Cantor's paradise in the safe turf of finitary reasoning. Most importantly, it uncovers the astonishing depth of Hilbert's philosophical thought. I present this fresh look at Hilbert's views in Chapter 2. In Chapter 6 I return to their appraisal and try to relocate Hilbert on the map of philosophers of mathematics. In the intervening chapters, I sketch the link between Hilbert's philosophical thought and the development of meta-mathematics. There, two morals surface. First, the formal science that Hilbert invented is neither an accidental by-product of some bad ideas nor a philosophically inert discipline whose conceptual origins are forever closed off from study. Rather, the principal techniques of meta-mathematics emerge directly from Hilbert's philosophical vision. Second, modern logic risks steering off a promising course if its practitioners lose sight of this fact. The flexibility of meta-mathematics continues to offer logicians ways to bring mathematical techniques to bear on questions about how and why those very techniques work, just as Hilbert proposed.

But before the recovery and development of Hilbert's views can commence, some of its prehistory is in order. This is the story about nineteenth-century mathematical creativity engendering the "epistemological crises" that led so many mathematicians to feel a need to espouse philosophical views about their science in the first place. The story is a largely familiar one, except that in the place of a single looming crisis I find several competing ones. The first step in understanding Hilbert is identifying the crisis that he was reacting to.

I.2 FREEDOM FROM NATURE

What today goes by the name "pure mathematics," the nineteenth-century German mathematician Georg Cantor called "free mathematics."[2] "Mathematics," he wrote, "is entirely free in its development and its concepts are restricted only by the necessity of being non-contradictory and coordinated to concepts previously introduced by previous definitions" ([1883], p. 896).

Prior to the nineteenth century,[3] from the advent of the scientific revolution, the disciplinary distinction between mathematical and empirical sciences familiar today was unknown. In the words of Penelope Maddy, "[t]he great thinkers of that time – from Descartes and Galileo to Huygens and Newton – did mathematics as science and science as mathematics without any effort to separate the two" ([2008], p. 17). In their able hands, it was a potent mixture. Both the physical and mathematical contributions that hindsight picks out from their work are formidable. But by the late 1800s, the mathematical vanguard considered any tendency to blur the line

---

[2] In [1883] Cantor emphasized the philosophical significance of his preferred terminology: He wrote that mathematicians are under "*absolutely no* obligation to examine their [ideas'] transient reality" and that "[b]ecause of this remarkable feature – which distinguishes mathematics from all other sciences and provides an explanation for the relatively easy and unconstrained manner with which one may operate with it – [mathematics] especially deserves the name of *free mathematics*, a designation which, if I had the choice, would be given precedence over the now usual 'pure' mathematics" (p. 896).

[3] The story of shifts in mathematical thought between the seventeenth and twentieth centuries has been told many times. My retelling of it follows Morris Kline's exemplary historical work in Chapters 41, 43, and 51 of *Mathematical Thought from Ancient to Modern Times*.

separating mathematical and empirical investigations ideologically crippling. To *these* great thinkers, the most exciting and promising mathematical novelties were predominantly ones that had no motivation or correlate in nature. Whiggish insistence from their conservative colleagues, like Fourier, that "[t]he profound study of nature is the most fertile source of mathematical discoveries," that "[t]he fundamental ideas are those which represent the natural happenings" was to them a nuisance.[4] Mathematicians had endured criticism directed against their growing registry of "unnatural" preoccupations – negative numbers, complex numbers, non-commutative quaternions, non-Euclidean and multi-dimensional spaces – until the sheer bulk of these inventions (and the fact that even their critics freely utilized them in their research) began to make that criticism sound tired and monotonous. According to Morris Kline, "after about 1850, the view that mathematics can introduce and deal with arbitrary concepts and theories that do not have any immediate physical interpretation ... gained acceptance" ([1972], p. 1031). To Cantor, as to many of the mathematicians most immersed in the farthest reaches of abstraction mathematics had to offer, this view was not only acceptable but definitive of their discipline's very nature. "The essence of mathematics," he wrote, "lies in its freedom" ([1883], p. 896).

With freedom from the restrictions of the empirical world came a license for unbridled creativity. Mathematicians came to see their work as an essentially creative activity. As a result, they began to see human genius not only as necessary in order to conquer mathematical terrain, but also as somehow constitutive of that terrain. Cantor's procession of extravagantly infinite sets was deservedly the

---

4 The passages are from the preface of Fourier's *Analytical Theory of Heat* as they are quoted in Kline [1972], pp. 1036–7.

centerpiece of this new scene, its intrigue deriving precisely from its fantastic, unnatural proportions. But the celebration of free creativity extended to every recess of mathematical activity. Of the laws of algebra, Alfred North Whitehead wrote: "They depend entirely on the conventions by which it is stated that certain modes of grouping the symbols are to be considered as identical" ([1898], p. 11). "It is obvious," he pointed out, "that we can take any marks we like and manipulate them according to any rule we choose to assign" (ibid., p. 4). His specific task was to disavow conservative algebraists of the idea that the algebraic laws depend in some way on arithmetical phenomena. Dramatically, Richard Dedekind had made a similar point regarding arithmetic itself. The preface to his monumental [1888] essay "Was sind und was sollen die Zahlen?" contains a representative disclaimer:

In speaking of arithmetic (algebra, analysis) as merely a part of logic I mean to imply that *I consider the number-concept entirely independent of the notions or intuitions of space and time* – that I rather consider it an immediate product of the pure laws of thought. My answer to the problems propounded in the title of this paper is, then, briefly this: *numbers are free creations of the human mind*; they serve as a means of apprehending more easily and more sharply the difference of things. ... [W]e are able accurately to investigate our notions of space and time by bringing them in relation with *this number domain created in our mind*. (p. 791, my emphases)

Decades earlier, in a letter he wrote to Bessel in 1811, Gauss urged that the totality of mathematics be seen this way. "One should never forget," he wrote, "that the functions [in complex analysis], like all mathematical constructions, are only our own creations, and that when the definition with which one begins ceases to make sense, one should not ask 'What is it?' but 'What is it convenient to assume in order that it remain significant?'" (quoted by Kline [1972], p. 1032). Even Hermann von Helmholtz, who, because of the methodological rift between his pioneering work in projective geometry and the more explicitly creative work of Riemann, is widely regarded as the century's most tenacious defender of the empirical nature of mathematics, described the structure of the natural number series as a human convention:

We may consider numbers initially to be a series of arbitrarily chosen symbols, for which we fix only a certain kind of succession as the lawlike or – as it is commonly put – the "natural" one. Its being termed the "natural" number series was probably connected merely with one specific application of numbering, namely the ascertaining of the cardinal number of given real things. [. . . But] their sequence too could be specified arbitrarily, so long as some or another specified sequence is immutably fixed as the normal or lawlike one. *This sequence is in fact a norm or law given by human beings, our forefathers, who elaborated the language.* I emphasize this distinction because the alleged "naturalness" of the number series is connected with an incomplete analysis of the concept of number. ([1887], pp. 730–1, my emphases)

The conception of mathematics as a creative human activity[5] became ubiquitous and deeply entrenched.

The Talmud records a debate about whether emancipation always benefits a slave (Gittin, 11b–13a). A slave knows where his next meal will come from; he knows where he will rest his head in the evening; he might even enjoy some of the seedier aspects of slave-culture from which a free man must abstain. On these grounds, Rabbi Meir reasons that freeing him can be detrimental. The Sages disagree: The obvious benefits of liberty outweigh the loss of security that Rabbi Meir cites. (Classical Jewish law follows the Sages' opinion.[6])

---

[5] Michael Detlefsen, whose discussions of these issues has greatly shaped my understanding of the era's mathematical thought, calls the orthodox nineteenth-century view of mathematical activity "creativist."

[6] One might wonder how this theoretical discussion has any legal consequences. In the Talmudic passage cited, the issue is whether a slave can be freed in his absence, through an

Mathematics' emancipation from empirical science presented the same dilemma. The attractions of unencumbered creativity shaping nineteenth-century mathematicians' research came at some apparent cost. As nature's slave, Enlightenment mathematics was secure. As empirically justified facts about the structure of the world, mathematical laws could not possibly admit contradictory results. Mathematical principles were abstract then, just as they are today, but as Locke influentially explained they were *abstracted from nature* so that the actual world, ever so subtly smoothed out, was an obvious model attesting to their consistency. But when one must live off one's wits, one takes one's fate into one's own hands. When he said that they were "restricted only by the necessity of being non-contradictory," in the same breath with which he otherwise liberated them, even Cantor placed one crucial condition on his colleagues' fanciful creations. The only problem was that, since consistency had always come for free, without so much as a thought, no one knew how to verify it of the new mathematics.[7] Unbridled creativity always flirts with the perils of incoherence, and mathematicians began their exodus from centuries of comfortable servitude with no principled way to guarantee they would stay out of trouble.

It did not take long for trouble to appear. In [1897] Cesare Burali-Forti discovered that the sequence of all ordinal numbers, which according to Cantor's set theory exists and is well-ordered, out-ranks in size all well-ordered sequences. But then the ordinal number corresponding to this sequence would, absurdly, be greater than all ordinal numbers, including itself. Kline ([1972], p. 1003) notes that Cantor had discovered this difficulty on his own two

---

agent. If freedom is beneficial, then emancipation *in absentia* is valid, for one may assume that the slave would not object to his release.

[7] No one knew how to verify it of the old mathematics, either, after scientists began to accept that even Euclidean geometry failed to describe the physical world.

years earlier. In letters to Dedekind in 1899,[8] Cantor pointed out an analogous difficulty lurking in the notion of the set of all cardinal numbers. The latter is essentially the contradiction that Bertrand Russell observed in the system that Gottlob Frege had designed in order to show that the laws of arithmetic could be derived from pure logic.

In short time, these "paradoxes" came to be seen not only as problematic for projects like Cantor's and Frege's, but as possibly symptomatic of the erroneous nature of constructions used throughout mathematics. One often cited example is the definition for least upper bounds. Their construction is "impredicative" – i.e., it defines a mathematical object in terms of the properties of a collection containing that object itself – and as such is analogous to the pathological constructions of set theory. Thus inconsistency threatened even seemingly well-established fields like real analysis.

1.3 FREEDOM FROM PHILOSOPHY

The discovery of the paradoxes ushered in a sense of panic. The idyllic era of free creativity appeared to have been founded on a mistake. By tearing mathematics aways from nature, mathematicians had uprooted it from the familiar source of its claim to truth. In place of truth, they focused on its newly found extravagance and increased ability to solve problems. Now this, too, was under fire by the threat that the whole enterprise might be inconsistent, that any notable theorem might be unhappily complemented by a contradictory one the next day. The ensuing panic inspired a thundering declaration from the mouths and pens of working mathematicians everywhere of an "epistemological crisis" facing mathematics.

---

[8] These letters are reprinted in Ewald [1996c], pp. 931–7.

The motivations behind the individual cries were quite different, though. Something, it was agreed, had to change, but there was no agreement about what features of mathematical practice were most threatened, worth preserving, and in need of defense.

To some thinkers, the paradoxes themselves did not engender crisis. They only made an existing one more apparent. Frege had hoped that by carrying out his derivation of arithmetic from logic, the rules of right-reason would fill the role that nature once filled. For him the crisis was inherently foundational. The recent development of mathematics made it evident, he felt, that the science could no longer be viewed as grounded in empirical observation. Frege mocked attempts by philosophers like Mill to maintain that it is. "What a mercy," he wrote in his *Grundlagen der Arithmetic*, "that not everything in the world is nailed down, for if it were" then Mill's attempt to define the number 3 in terms of the rearrangement of physical things would cease to have any meaning, "and $2 + 1$ would not be 3"(§7). But Frege worried that mathematics' newly found freedom from the confines of nature had fueled an unhealthy, subjectivist view of its underpinnings. Thus he strove to show that mathematics flowed, not from the haphazard patterns of thought that humans chose to pursue – a view he labeled "psychologism" – but from the timeless and impersonal constraints on how they *ought* to reason, from the pure laws of logic. "Otherwise," he wrote, "in proving Pythagoras' theorem we should be reduced to allowing for the phosphorous content of the human brain" (Introduction p. vi). He concluded:

Weird and wonderful are the results of taking seriously the suggestion that number is an idea. And we are driven to the conclusion that number is neither spatial and physical, like Mill's piles of pebbles and gingersnaps, nor yet subjective like ideas, but non-sensible and objective. Now objectivity cannot, of course, be based on any sense-impression, which as an affection of our mind is entirely subjective, but only, so far as I can see, on the reason. (§27)

Frege did caution that without a thorough and proper grounding of mathematics like what he claimed to offer, mathematics was not properly rigorous, its theorems not adequately demonstrated, and its principles not necessarily consistent:

[I]t must still be borne in mind that the rigor of the proof remains an illusion, even though no link be missing in the chain of our deductions, so long as the definitions are justified only as an afterthought, by our failing to come across any contradiction. . . . and we must really face the possibility that we may still in the end encounter a contradiction which brings the whole edifice down in ruins. (ix)

But the threat of paradox itself was not the crisis he was reacting to. There is no evidence that he foresaw contradiction as a living possibility. Frege was troubled by the very presence of what he considered ignoble conceptions of mathematics. He pointed to the (theoretical) possibility of the mathematical edifice toppling down, and to his peers' inability to explain why such a disaster could not happen, in order to expose the inadequacy of their views as accurate depictions of the nature of mathematics. A correct account of mathematical activity, he felt, should make its consistency evident. Frege believed that mathematics was a pure science, free from "psychological influences" and "external aids," that "it is precisely in this respect that mathematics aspires to surpass all other sciences, even philosophy" (iii–iv). Improper grounding jeopardizes mathematics' reputation, soils it by dragging it through Mill's muck or worse, in psychologism, "makes everything subjective, and . . . does away with truth" (vii). "I found myself forced to enter a little into psychology," he wrote, "if only to repel its invasion of mathematics" (viii). Frege's crisis was the threat of mathematics falling into ill repute. His agenda was to defend its nobility.[9]

But to the dismay of visionaries like Frege, most mathematicians were untroubled by the groundlessness of mathematical activity until they learned of the paradoxes. The legend about mathematics being securely grounded in nature before the advent of infinitesimals and non-Euclidean geometry, about how Enlightenment mathematicians never worried over the consistency of their methods because the real world was their obvious model, is largely anachronistic. In those days, the question of the consistency of mathematical principles simply did not arise. The idea of nature, or anything else, grounding their science, in the sense that has become fashionable to speak about, was just as foreign to them as the idea of their results being inconsistent. The generation of mathematicians who worked after the division of mathematics and empirical science were thus largely untroubled by any foundational crisis. They were concerned with concocting ever more fanciful creations for ever more ingenious solutions to mathematical problems, and they were confident that success along these lines would speak more for the nobility of their craft than would any philosophical doctrine about tapping into a third realm beyond mind and matter.

On this majority view, only a literal threat of inconsistency could inspire the kind of foundational research that Frege called for. But the perception of any such threat facing mainstream mathematical activity was minimal for most of the nineteenth century. Frege called his own investigations "more philosophical than many mathematicians may approve" and described those mathematicians'

---

[9] Like Frege, Kronecker also sensed epistemological crisis prior to the discovery of the paradoxes. The crisis he perceived was inherently foundational in much the same way Frege's was. After the discovery of the paradoxes, the Intuitionists took refuge in the framework he set up, just as the logicists tried to revive Frege's.

glibness as a "scandal," not because of any heightened sense of caution on his part, but because inconsistency was not his chief concern (v). The set-theoretical paradoxes irreparably closed this ideological gap between Frege and the bulk of the mathematical world. Reluctantly, mathematicians took to foundational investigations in order to ward off any further calamity. *Their* crisis was the very real concern that the triumphs of modern mathematics would be swept away one after another, as the edifice continued to crumble.

Although his disclosure to Frege of the paradoxes and their devastating effect on his program led Frege to despair of its prospects, Russell himself took up the "logicist" project, amending certain aspects of Frege's system so as to avoid paradoxical constructions. The Intuitionists, led by L. E. J. Brouwer, embraced a conception of mathematics as a mental activity. To ward off the threat that inconsistency might topple the mathematical edifice, they began preemptively carving away from it all principles and results that do not correspond with a certain type of mental construction. They believed they could evade the charge of subjectivism by invoking certain features of Kant's transcendental philosophy. Herman Weyl, following Henri Poincaré, proposed a similar technical reform, through a prohibition of all impredicative definitions. He backed this up with a semantic theory of properties inspired by Husserl's phenomenology.[10]

Thus the later foundationalists differed from their predecessors in that they were trying to secure (at least some portion of) actual

---

[10] Carnap's [1931] essay "The logicist foundation of mathematics" remains a valuable introduction to the logicist program. The introductory essays in Mancosu [1998a] by Walter van Stigt and Paolo Mancosu on Brouwer and Weyl, respectively, contain particularly helpful explanations of the basic ideas behind Intuitionism and Predicativism, as well as their historical contexts.

mathematics by grounding it in an unassailable form of reasoning or mental construction. This motive proved to be a good deal more attractive than the perceived need to defend mathematics' reputation. Mathematicians who had been unmoved by Frege's worries now flocked to one or another of the foundational schools. But the feeling that mathematics needed grounding in logic (or Kantian intuition, or Husserlian meaningfulness) was not universal. A few thinkers found the sudden interest in foundationalism deeply confused and threatening on its own.

Independently of Frege, the American polymath Charles Sanders Peirce had devised a great deal of modern propositional and quantificational logic in the last decades of the nineteenth century.[11] His father, Benjamin Peirce, was a notable algebraist and early enthusiast of Hamilton's quaternions. William Ewald describes the senior Peirce's generalization of Hamilton's work as "producing a bounty of algebraic structures with unusual multiplication tables" ([1996a], p. 597). Peirce followed his father's mathematical research closely, often discussing with him algebraic novelties and the nature of mathematics. His own work in logic was squarely in the algebraic tradition, as is evident in the title of his most significant logical paper, "On the algebra of logic."

Characteristic of Peirce's logical investigations is his eagerness to bring the whole gamut of modern mathematical novelties to bear on his subject. He saw in his father's wildest creations opportunities to codify and elucidate the behavior and interaction of logical connectives and reasoning patterns. So eager to deploy the most sophisticated mathematical techniques was Peirce, that his [1870] paper "Description of a notation for the logic of relatives," in which

---

[11] I am indebted to Cornelius Delaney for conversations about the relationship between the thought of Peirce and Hilbert.

he extended Boole's algebraic treatment of monadic predicates to multi-place relations is, according to Ewald, "filled with logarithms, exponentiations, and power series expansions whose logical significance is difficult to fathom" ([1996a], p. 597). In short, Peirce saw in mathematics, if not a foundation for logic, at least the tools to develop its science.

The reaction of most mathematicians and philosophers to the paradoxes of set theory thus seemed to Peirce exactly backward. The scientific investigation of logic was in its formative stages. The rules of right reasoning still were unknown. A mature science like mathematics, with a history of successful elucidation and problem solving, was needed in order to develop logic. Peirce believed that difficulties within mathematics, like those that had appeared in set theory, would probably be solved by mathematicians. He was certain that mathematics develops through its confrontation with and solution of problems, and he embraced the idea that human creativity unencumbered by fixed rules of inquiry is an essential ingredient in that development. The use of logic to prune and shape mathematical activity – the idea that logic might be the ground of mathematical activity – struck Peirce as absurd. "In truth," he wrote, "no two things could be more different than the cast of mind of a logician and that of a mathematician. It is almost inconceivable that a man should be great in both ways" ([1876], p. 595). He thought the idea that a faculty of the human intellect or intuition might be mathematics' foundation exemplified an unhealthy trend to approach philosophy mathematically. He had in mind the logicists specifically when he wrote:

One singular consequence of the notion which prevailed during the greater part of the history of philosophy, that metaphysical reasoning ought to be similar to mathematics, only more so, has been that sundry mathematicians have thought themselves, as mathematicians, qualified to discuss philosophy; and no worse metaphysics than theirs is to be found. ([1902], p. 639)

The seeds of Peirce's insistence on the priority of mathematics over logic can be found already in his father's work. In [1870] Benjamin Peirce, using "law" to refer to laws of nature and reason, wrote that "neither law can rule nor theory explain without the sanction of mathematics," for "it is the judge over both, and it is the arbiter to which each must refer its claims" (p. 585). These words echo in the writing of his son, who, arguing again against the logicists, wrote:

It does not seem to me that mathematics depends in any way upon logic. It reasons, of course. But if the mathematician ever hesitates or errs in his reasoning, logic cannot come to his aid. He would be far more liable to commit similar as well as other errors there. On the contrary, I am persuaded that logic cannot possibly attain the solution of its problems without the great use of mathematics. Indeed, all formal logic is merely mathematics applied to logic. ([1902], p. 638)

To Peirce, and to his intellectual heir John Dewey,[12] the paradoxes of set theory posed not a crisis, but a problem internal to mathematics. The epistemological crisis he sensed was the failure of mathematicians to see this, as their heretical turn from a pragmatic to a foundational view of their craft made evident. Mathematics risks being radically distorted if it is shaped according to the dictates of a fledgling science instead of by the continuous confrontation with and creative solution of ordinary mathematical problems.

Back in Europe, Hilbert was not particularly concerned to defend any particular conception of the nature of mathematics. In this way he differed from Frege and Peirce. But like Peirce, and unlike the

---

[12] Sidney Ratner's [1992] essay on Dewey's philosophy of mathematics traces Peirce's ideas through Dewey to their revival in the late twentieth century.

post-paradox foundationalists, he was equally unconcerned with a perceived groundlessness of recent mathematical activity. Hilbert saw how modern mathematics blossomed once it was no longer tied down to nature. Its richness, he believed, derived from the very fact that mathematicians no longer had to cite anything external to mathematics to justify their creations. Now this sense of freedom was threatened, not by the paradoxes in set theory, but by the mathematical world's reaction to them.

It did not matter much to Hilbert whether or how much of Cantor's theory of sets could be grounded in the laws of reason, Kantian intuition, or the self-correcting mechanisms of the pursuit of truth. What mattered was that Cantor developed set theory without having to appeal to these doctrines, that the need to do so would have crippled his efforts. The era of free creativity in mathematics was vindicated in Cantor's development of what Hilbert called "the finest product of mathematical genius and one of the supreme achievements of purely intellectual human activity" ([1926], p. 188). Philosophical encroachment was unwelcome: Under its guidance, mathematics would only find itself re-enslaved, serving philosophy as it had once served nature. "The philosophers," he wrote, "will be interested that a science like mathematics exists at all." They will be puzzled by how human creativity can successfully solve problems and advance knowledge, and they will try to reign it in in order to dispel that mystery. "For us mathematicians, the task is to guard it like a relic" ([1931], p. 273). Quite naturally, Hilbert believed that guarding mathematics involves more than retaining its current results. It requires preserving the conditions under which mathematical activity can continue to flourish.

"No one," Hilbert proclaimed, "shall drive us out of the paradise which Cantor has created for us" ([1926], p. 191). Cantor's paradise begins with his theory of sets. It extends outward to the environment of unencumbered pursuit of new mathematical creations that allowed a vision like Cantor's to materialize. Hilbert designed his theory of proofs in order to defend that whole realm. Indeed, it was the spirit of mathematical research and the freedom that its practitioners enjoyed that he most hoped his new science would preserve. The first stage of that program was the "axiomatic method," which, he wrote,

is and remains the one suitable and indispensable aid to the spirit of every exact investigation no matter what domain; it is largely unassailable and at the same time fruitful; it guarantees thereby *complete freedom of investigation*. ([1922], p. 201, my emphases)

Although Hilbert recognized a philosophical assault on mathematical freedom and creativity, he chose not to fight back. A better way to fend off the philosophers, he realized, is to make them lose interest in their projects. He aimed to smother the foundational urge.

A unique characteristic of Hilbert's thought is that the question of the priority of logic over mathematics does not arise. The logicists, and to some extent other foundationalist thinkers, wanted to rebuild mathematics on logic, thereby infusing mathematical techniques with the security of logical conviction. Against this suggestion, Peirce took mathematics to be prior. Hilbert saw through both sides of this debate:

Arithmetic is often considered to be a part of logic, and the traditional fundamental logical notions are usually presupposed when it is a question of establishing a foundation for arithmetic. If we observe attentively, however, we realize that in the traditional exposition of the laws of logic certain fundamental arithmetic notions are already used ... ([1904], p. 131)

The feeling of paradox is therefore inevitable as soon as one seeks justificatory grounds in this arena. Hilbert proposed to shake this feeling by giving up on justification altogether. He continued: "Thus we find ourselves turning in a circle, and that is why a partly simultaneous development of the laws of logic and of arithmetic is required if paradoxes are to be avoided" (ibid.). In the 1920s "the simultaneous development of logic and mathematics" became one of Hilbert's favorite slogans.[13] By accepting the fact that logic and mathematics are intertwined, Hilbert deliberately stripped from each the illusion that they are rooted in the other. He did this, not to show that they are unfounded, but to show that they do not call for foundation. "If mathematical thinking is defective," he wrote, "where are we to find truth and certitude?" ([1926], p. 191).

Hilbert did not then turn to this logico-mathematical amalgam and seek a foundation for it. Instead, he set out to rid the world of the temptation to ground an activity that thrives on freedom. This he believed he had done, not by arguing against the idea of foundationalism, but by supplanting it with a mathematical alternative. In [1931], as meta-mathematics was nearing maturity, he wrote: "I believe that in my proof theory I have fully attained what I desired and promised: The world has thereby been rid, once and for all, of the question of the foundations of mathematics as such" (p. 273).

Remarkably, Hilbert thought that all the spoils sought by champions of competing causes would follow from his banishment of

---

[13] An example is Hilbert's reflection in [1922] on the technical challenge of formalizing "object level" mathematical theories, the formalalization of which leads to meta-mathematical investigation: "This circumstance," he wrote, "corresponds to a conviction I have long maintained, namely, that a simultaneous construction of arithmetic and formal logic is necessary because of the close connection and inseparability of arithmetical and logical truths" (p. 211). Hilbert's "simultaneity" slogan has been unpacked in various technical expositions, notably Richard Zach's [2003]. The philosophical point that Hilbert was making has gone unnoticed, though. Here, Hilbert dates his conviction that neither logic nor math can ground the other to 1900.

philosophy from mathematics. By preserving mathematicians' freedom, both the mathematical edifice and its nobility would be preserved. For he did not intend merely to persuade other mathematicians to resume their accustomed ways – he aimed to show mathematically that those ways were unproblematic. For this reason he spoke often and eloquently about his efforts to re-instill mathematics' "old reputation of incontestable truth" as well as its inventory of profound results. But those prizes were, to Hilbert, worth little on their own, without the license for mathematicians to continue to extend freely into new areas of thought. That license was his chief aim.

1.4 THE WRONG CONCLUSION

In Section 1.1 I proposed that a reasonable, pragmatic approach to the history of ideas should lead one to posit that Hilbert's philosophical thought is worth serious attention simply because the science he invented is so revolutionary. Not everyone will agree that such historical hypotheses are always warranted, but I hope that reviewing Hilbert's "anti-philosophical" position has removed one obstacle to approaching his thought with optimism. I am not proposing that Hilbert's ideas about the nature of mathematical phenomena or his ideas about our ability to know about these phenomena are justified by the fact that they somehow inspired him to forge his new science. Since ideas of that sort might very well inspire scientific innovations without thereby being vindicated, the suggestion that they should be re-evaluated in light of their contributions to culture is not particularly persuasive.[14] Thus the inability to

---

[14] In *Naturalism in Mathematics*, drawing on examples from Albert Einstein's use of philosophical principles, Maddy distinguishes cases where philosophy is tied up in "the justificatory structure of scientific practice" and when it serves as mere inspiration. My

appreciate Hilbert's thought results from the assumption that his philosophical views were such idle ruminations about the nature of mathematics. This, in turn, results from the assumption that underlying Hilbert's rejection of Frege, Brouwer, and Weyl's attempts to provide a philosophical grounding for mathematics was some alternative foundational position that Hilbert preferred. But this whole train of assumptions misses Hilbert's main idea, which is that mathematics does not need philosophical foundations any more than it needs to be grounded in nature.

It is tempting to conclude that Hilbert's vision for his new science differed very little from the view of it that Tarski would soon adopt. It is tempting to see these two mathematicians in full agreement that meta-mathematics is just more mathematics and that philosophical considerations have no bearing on its history or development. The only real difference between the two, one might think, is that Tarski mistook Hilbert to be engaged in a philosophical project, whereas Hilbert knew full well that he was doing just the opposite. But giving in to this temptation would be to misunderstand Hilbert all over again. Hilbert did not have a "philosophy of mathematics" in the sense that Plato, Kant, and Frege had theirs. He had no views about what mathematical activity "consists in," for he thought that all such views endanger mathematics. But this does not mean that he had no philosophical views at all.

Hilbert believed that science in general, and mathematical activity in particular, can advance human understanding both of the world we live in and of the process of understanding itself. His disinterest in traditionally philosophical projects was based not on a

---

suggestion is that Hilbert's thought was not of the latter sort, but rather figures in his mathematical development in ways that Maddy suggests allow it to be validated along with the mathematics. See Maddy [1997], pp. 188–93. See the quotation on p. 87 of this book for a similar idea in the writing of Jacques Herbrand.

dismissal of those projects' goals, but on a rejection of the idea that attaining them calls for extraordinary, philosophical means. Hilbert believed that ordinary, "first-order" inquiry – scientific and artistic activity – should suffice. Thus he described Cantor's mathematical work as providing "the deepest insight into the nature of the infinite," and he called set theory "a discipline which comes closer to a general philosophical way of thinking [than analysis] and which was designed to cast new light on the whole complex of questions about the infinite" ([1926], p. 188). But he did not agree with Cantor that "even for finite multiplicities a 'proof' of their 'consistency' cannot be given," that "the fact of the 'consistency' of finite multiplicities is a simple, unprovable truth" (Cantor [1899], p. 937). Before Cantor, the infinite was philosophical terrain; before Hilbert, knowledge of the consistency of mathematical principles was. Designs to annex philosophical territory through artistic or scientific revolutions are not signs of philosophical naiveté or naturalistic fallacy, but of genius.

Tarski called Hilbert's thought "theology." When Hilbert described his investigations, he characterized them in exactly the opposite way:

Already at this time I should like to assert what the final outcome will be: mathematics is a presuppositionless science. To found it I do not need God, as does Kronecker, or the assumption of a special faculty of our understanding attuned to the principle of mathematical induction, as does Poincaré, or the primal intuition of Brouwer, or, finally, as do Russell and Whitehead, axioms of infinity, reducibility, or completeness ... ([1928], p. 479)

As always, Hilbert does not argue for his claim. He was not much given to argument, preferring instead just to say what he thought and to let tomorrow's results attest to the merits and faults of his ideas. He believed that the fact that mathematics does not need grounding could only become evident through the development of mathematics itself. Not having a view about the nature of mathematics is "anti-philosophical," so far as it goes. But maintaining that no such doctrine is necessary is a philosophical view of its own – one that set Hilbert's thought apart from his peers'. In place of views about the nature of his craft, he had unrivaled philosophical insight about its potential uses.

History has not been kind to Hilbert in its depiction of him turning from mathematics to philosophy to find reassurance that mathematical techniques are justified. By replacing that image with another one – of Hilbert turning mathematics against his colleagues' urges to find a foundation for it – I hope that I have salvaged Hilbert's vision of a philosophical idea essentially intertwined with a scientific program. It is this philosophical idea that I claim deserves to be recovered and re-evaluated in light of the success of Hilbert's new science.

CHAPTER 2

David Hilbert's naturalism

2.1 INTRODUCTION

In [1922] David Hilbert delivered a series of lectures to the Hamburg University Mathematics Seminar outlining the methods for a program that would occupy him and his colleagues for roughly a decade. His report repeatedly emphasized the role that a certain type of consistency proof was to play in the realization of the program's aims. After the mathematical and philosophical communities came to understand the theorems of Gödel [1931] as demonstrating the unavailability of this type of consistency proof, they rejected on that ground Hilbert's program as a failed attempt at securing the foundations of mathematics. Indeed Hilbert's program has come to be *identified* in many minds with the production of an appropriate sort of consistency proof – so much so that his contributions to the foundations of mathematics are simply, in the opinions of many, a glib "formalism" regarding the nature of mathematics, a somewhat vague, proto-constructivist "finitism" regarding mathematical existence, and a radical, "all eggs in one basket" thesis according to which mathematics is a terminally unfounded enterprise unless the consistency of a significant portion of it can be proven (impossibly) according to the demands of these two restrictive "isms."

This construal of Hilbert's program is difficult to reconcile with a continuous theme in his Hamburg lectures according to which one has every reason to believe in the consistency of mathematics, because of the clarificatory gains in the axiomatics of Weierstrass, Frege, Dedekind, Zermelo, and Russell and, most of all, because everything in our mathematical experience speaks for its consistency. If this sentiment amounts only to an optimism that the carefully described consistency proof will eventually be carried through, then it is a remarkably cavalier sentiment owing to the fact that the proof called for was an open problem in a completely new and largely uncharted field of logic. Moreover there is no hint as to why the clarification gained through axiomatics or especially through "mathematical experience" should weigh in on the prospects for such a proof. Instead Hilbert appears to be articulating in the report a pre-theoretical belief in the consistency of mathematics that on the one hand does not ride on the promise of a realization of his foundational program, and on the other hand carries on unscathed by announcements of skepticism from the philosophical schools. This is hardly the mindset of one who sees the question of the consistency of mathematics depending entirely on the existence of a proof radically constrained by two philosophical scruples.

Even more difficult to reconcile with the received view of Hilbert's foundational contributions is his announcement that he intends his treatment, which he describes as epistemological, to probe more deeply than previous investigations in the foundations of mathematics. The technical component of his program no doubt was a tremendous move forward in mathematical sophistication. The report introduced for the first time the notion of meta-mathematics and the prescription of separate techniques for the objectual and meta-mathematical levels of investigation as well as – through the introduction of proof theory – a specific way to carry out meta-mathematical investigations. But if the philosophical position inspiring these technical achievements is merely the belief that mathematics consists ultimately of meaningless formulas and that the valid inferential moves regarding these are the purely finitary ones, then neither does Hilbert's position seem deeper than any other nor does it appear to be answering deeper, or even different, concerns than those that troubled Frege, Brouwer, and Weyl. Hilbert must, instead, be advancing a program of considerable philosophical insight if he takes himself to be advancing "a deeper treatment of the problem."

In fact Hilbert's epistemological position differs significantly from those of his intellectual adversaries. Foundational concerns had traditionally fallen directly out of skepticism concerning questions like whether mathematics is consistent, and the rigor and methodology of the investigations undertaken by Frege, Brouwer, and Weyl were responsible precisely to the nature and degree of their own skepticism. By contrast, Hilbert is less skeptical than they are concerning the consistency of mathematics but at the same time has much higher standards for what counts as a proof of it. This is because the question inspiring him to foundational research is not whether mathematics is consistent, but rather whether or not mathematics can stand on its own – no more in need of philosophically loaded defense than endangered by philosophically loaded skepticism. All the traditional "Hilbertian theses" – formalism, finitism, the essential role of a special proof of consistency – are methodological principles necessitated by this one question. When they are understood in that light, they appear no longer to be the glib scientistic principles of an expert mathematician's amateur dabbling in philosophy. They appear rather to be the constraints on method needed for probing a deep epistemological issue left untouched by rival programs.

If Hilbert's program is understood thus, the temptation to ignore it as a contribution to the philosophy of mathematics goes away.

Hilbert's epistemological stance turns out to be one of philosophical subtlety and originality. Additionally, since it has been largely ignored in favor of a philosophically naive, purely mathematical program, the viability of the program in the light of the development of logic in the last thirty-five years remains largely unexplored. I hope to show here that Hilbert's program *vis-à-vis* mathematical autonomy is philosophically instructive.

<h2 style="text-align:center">2.2 ANTI-FOUNDATIONALISM</h2>

The principal concern leading mathematicians and philosophers to consider the foundations of mathematics in the early twentieth century was the discovery of the paradoxes in nineteenth-century set theory and in Frege's axiomatic system. These systems were designed to provide a conceptual framework for significant portions of mathematics in response to points of unclarity their designers sought to eliminate. Since the paradoxes played the ironic move of slipping into notice precisely at the point of purported certitude, they were thought of as ushering in an epistemological crisis. The paradoxical nature of mathematics had been chased directly to the conceptual framework on which the science was thought to rest. Since this framework itself had proven untenable, the new foundational task was to provide a replacement. The project was daunting. On the one hand, since mathematics was to be secured on a framework other than the one on which mathematicians had been basing their methods, there was no guarantee that the new, well-founded mathematics would resemble even closely the old. In addition, entirely new methods needed to be developed, both to build mathematics up from whatever new framework was decided on and to practice the science according to this framework. On the other hand, the prospects for deciding quickly on a new framework were dim. The decision amounted to selecting an epistemological theory – the *correct* epistemological theory for mathematics – and the only unanimously accepted and scientifically informed principle of theory selection was to reject paradoxical ones. This opened wide the foundational search to reflection on the nature of mathematics, consideration of the relationship between mental representation and the world, and other matters on which there was almost no consensus and equally little argumentative standard for how consensus might be reached.

Hilbert's reaction to the paradoxes and ensuing threat of mathematical inconsistency differs radically from the general response just described, but the exact nature of his reaction is somewhat elusive. His addresses routinely cycle back and forth between announcements such as this one of literal uncertainty in the consistency of mathematical theories: "we can never be certain in advance of the consistency of our axioms if we do not have a special proof of it" ([1922], p. 201), and statements like the following of his utmost certainty in the validity of mathematical methods: "the paradoxes of set theory cannot be regarded as proving that the concept of a set of integers leads to a contradiction" since "[o]n the contrary, all our mathematical experience speaks for the correctness and consistency of this concept" ([1922], p. 199). This seeming tendency to second guess or contradict himself suggests that Hilbert had no fully developed view either of the impact of the paradoxes or of the state of foundations.

One way to resolve this tension is to attribute both Hilbert's call for a concrete proof of consistency and his claim that the consistency of mathematics depends entirely on the existence of such a proof to his considered philosophical position, while writing off his declarations of assurance in "the correctness and consistency of mathematics" as mere academic optimism that in due time such a proof would surface from his research circle. So prone to organizing his addresses around similar optimistic proclamations (the infamous call to arms against the "*ignorabimus*" in mathematics and the suggestion that mathematics was destined to subsume within its scope all of human knowledge are two glaring examples) was Hilbert, that this picture would seem at first not at all unreasonable. And consistently with this reading one could place Hilbert's foundational philosophy alongside his contemporaries'. He would, in particular, be in agreement with them that the discovery of paradox in current foundational programs justifies skepticism in the consistency of mathematics which can only be answered by re-securing mathematics on new foundations. Distinguishing Hilbert's proposal would be only his insistence that the absence of paradox in the new foundations should be *mathematically proven* rather than justified somehow *a priori*, and the optimism – perhaps inspired by Hilbert's distinguished mathematical tenure – that all orthodox mathematics could in this way be grounded. The passage most suggestive of this reading is the following one from the 1931 article in *Mathematische Annalen* where Hilbert characterizes his belief in the consistency of mathematics as "faith" and proceeds to claim that "faith," in this case, does not suffice:

It would be the death of all science and the end of all progress if we could not even allow such laws as those of elementary arithmetic to count as truths. Nevertheless, even today Kronecker still has his followers who do not believe in the admissibility of *tertium non datur*: this is probably the crassest lack of faith that can be met with in the history of mankind.

However, a science like mathematics must not rely upon faith, however strong that faith might be; it has rather the duty to provide complete clarity. ([1931], p. 268)

There is another way to resolve the tension, however, that puts Hilbert directly at odds with his contemporaries' epistemological views. Unlike the attempt just described, moreover, this resolution is in keeping with Hilbert's opening statement in his Hamburg lectures:

If I now believe a *deeper treatment of the problem* to be requisite, and if I attempt such a deeper treatment, this is done *not so much to fortify individual mathematical theories* as because, in my opinion, all previous investigations into the foundations of mathematics fail to show us a way of *formulating the questions concerning foundations so that an unambiguous answer must result*. But this is what I require: in mathematical matters there should be in principle no doubt; it should not be possible for half-truths or truths of fundamentally different sorts to exist. ([1922], p. 198, emphasis added)

That is, Hilbert deliberately intends a deeper foundational investigation than those of his contemporaries, and his chief aim in doing so is not to demonstrate the consistency of any branch of mathematics. It is, rather, to establish a mathematical autonomy according to which the reliability and correctness of ordinary mathematical methods does not rest on *any* epistemological background – neither the failed conceptual framework of nineteenth-century set theory, nor any new philosophically informed framework – since these can only ever provide "ambiguous" foundations – foundations dependent in their conclusiveness on their underlying philosophical principles. Since philosophical principles are, according to Hilbert, eternally contentious, such a defense of mathematics would only be a "half-truth": a truth only in so far as one is willing to subscribe to the relevant philosophical principles. Hilbert shares his adversaries' goal: "The goal of finding a secure foundation of mathematics is also my own ... " But finding secure foundations is for Hilbert just to cut through the fog of such half-truths: "I should like to regain for mathematics the old reputation of *incontestable* truth" since this reputation for objectivity more than anything else is that "which [mathematics] appears to have lost as a result of the paradoxes of set theory" ([1922], p. 200, emphasis added).

The apparent conflict between Hilbert's determined affirmation of the consistency of mathematics and his desperate call for a proof of it is properly settled in precisely the opposite manner of the received view: His considered philosophical position is that the validity of mathematical methods is immune to all philosophical skepticism and therefore *not even up for debate* on such grounds. The consistency proof of a precisely delineated sort is just a methodological tool designed to get everyone, unambiguously, to see this.

On what grounds might one reasonably retreat to skepticism about the veracity of mathematical methods? To recognize a contradiction in a mathematical system is straightforward once it has been discovered. For example one is able to verify Russell's paradox in Frege's system directly, and so skepticism about the veracity of the system in Frege's *Grundgesetze* is perfectly reasonable and insurmountable. It is questionable, though, how one might foster this same sort of skepticism about a system for which one cannot formally demonstrate any inconsistency. Presumably one would need to be reasoning from some quite elementary standpoint, the security of which one takes to be granted for present purposes but from which mathematical methodology seems both in need of justification and under threat of instability. One option would be to articulate some non-mathematical standpoint from which the system indeed does appear to be plagued by a noxious failing. If one could successfully defend the epistemological security of this standpoint, say by showing that it accords with appropriately elementary principles of reason and is therefore sounder than the mathematical system one is attacking from it, then again skepticism seems a reasonable refuge.

Such was the attack that the Intuitionists and Predicativists waged on orthodox mathematical practice. Hilbert's reaction to their epistemology is illuminating, for he directly challenges, not the epistemological security of the Intuitionist or Predicativist standpoints, but the general strategy just described. In fact, Hilbert does not distinguish the Predicativist and Intuitionist positions in his lectures, though his particularly detailed remarks about the charge of circularity is more applicable to Predicativism. Therefore it will be most appropriate to refer to the view he is opposing as the Predicativist view. Ultimately, however, since Hilbert criticizes the general method of skeptical foundationalism and not the details of the Predicativist position, the same critique applies equally well to the Intuitionist and any other similarly conceived program.

Just one year before Hilbert's Hamburg lectures, Weyl published a report "On the new foundational crisis of mathematics" outlining precisely the understanding of the impact of the set-theoretical paradoxes described above, as well as his attempt, and another due to Brouwer which Weyl had recently embraced, at re-centering mathematical practice on new, philosophically informed foundations. The introductory comments to this report are the primary philosophical target of Hilbert's address:

The antinomies of set theory are usually treated as border conflicts concerning only the most remote provinces of the mathematical realm, and in no way endangering the inner soundness and security of the realm and its proper core provinces. The statements on these disturbances of the peace that authoritative sources have given (with the intention to deny or to mediate) mostly do not have the character of a conviction born out of thoroughly investigated evidence that rests firmly on itself. Rather, they belong to the sort of one-half to three-quarters honest attempts of self-delusion that are so common in political and philosophical thought. Indeed, any sincere and honest reflection has to lead to the conclusion that these inadequacies in the border provinces of mathematics must be counted as symptoms. They reveal what is hidden by the outwardly shining and frictionless operation in the center: namely, *inner instability of the foundations on which the empire is constructed*. (Weyl [1921], p. 86)[1]

It is interesting that Weyl here charges the defenders of traditional mathematical techniques with trading in half-truths and self-deception, since Hilbert raises a very similar complaint with the idea of grounding mathematics on *a priori* principles. These charges seem indicative of research programs operating under fundamentally different conceptions of adequacy in scientific foundations. Weyl's main point is nonetheless clear: He takes himself to have identified the source of the antinomies, not in the experimental far reaches of mathematics, but in basic principles like impredicative definitions. Since classical mathematics abounds with such techniques even in its core research areas, Weyl took this circularity to undermine totally the veracity of ordinary mathematical practice.

Hilbert begins his objection to Weyl's skepticism by noting the artificiality of Weyl's standpoint:

[O]ne sees that for the mathematician various methodological standpoints exist side by side. The standpoint that Weyl chooses and from which he exhibits his vicious circle is not at all one of these standpoints; instead it seems to me to be artificially concocted. ([1922], p. 199)

Specifically, Hilbert criticizes Weyl's argument for resting on patently non-mathematical grounds. Immediately one wonders at the relevance of Hilbert's complaint. The standpoint of Weyl's criticism is non-mathematical, but couldn't it be reliable all the same? And if it is reliable, then should not the circularities demonstrable within it impugn classical mathematics? Moreover, what is one to make of Hilbert's charge of artificiality? Elsewhere he repeatedly emphasizes that mathematical systems themselves are fully

---

[1] Added italics follow Ewald's translation of Hilbert [1922], where this passage is quoted.

arbitrary, that they earn their credence simply by virtue of their consistency. Why, then, should the artificiality of Weyl's standpoint implicate it?

Hilbert's elaboration of his criticism is extraordinary:

Weyl justifies his peculiar standpoint by saying that it preserves the principle of constructivity, but in my opinion precisely because it ends with a circle he should have realized that his standpoint (and therefore the principle of constructivity as he conceives it and applies it) is not usable, that it blocks the path to analysis. ([1922], p. 199)

Hilbert rejects Weyl's standpoint and the philosophical principles behind it *because* of the circularity that from this standpoint appears in classical mathematics. There is no discussion of the reliability of Weyl's constructivism, no analysis of the degree to which the Predicativist standpoint is epistemically secure or elementary. Neither does Hilbert propose an alternative to Predicativism or explain where he thinks its philosophical underpinnings go wrong. Weyl's foundational program fails, in Hilbert's estimation, simply because "it blocks the path to analysis," because classical mathematics is not recoverable in it.

By the artificiality of a standpoint, then, Hilbert means that it is not native to mathematical practice. Whatever can be said for such a standpoint must in some way betray mathematical standards and therefore the mathematician is under no obligation to his science to take heed. The same year Bernays expressed the point as follows: "Thus we find ourselves in a great predicament: the most successful, most elegant, and most established modes of inference ought to be abandoned just because, from a specific standpoint, one has no grounds for them" ([1922b], p. 218). In such a predicament, there is only one way to turn, as Hilbert memorably explains in his celebrated [1926] address to the Westphalian Mathematical Society, "no one, though he speak with the tongue of angels, could keep people from negating general statements, or from forming partial judgments, or from using *tertium non datur*" because these principles are the mathematician's fundamental resources and arguments against them simply are of no weight next to our compulsion to work with them and the achievements attainable by them.

In short, since "[t]he standpoints usually taken by mathematicians do not rest on the principle of constructivity at all, nor do they exhibit Weyl's circle" ([1922], p. 199), the Predicativist must entice us to jump ship, to opt for the skeptic's subtle philosophy over mathematical methodology. But if from the mathematical mode of thinking nothing seems out of line, then the skeptic's call is just so much rhetorical sport and we are destined as a matter of fact to ignore it in favor of the clarity and naturalness of our science. One is reminded of Descartes' reaction at the end of his First Meditation to his own skeptical tendencies:

> But this undertaking is arduous, and a certain laziness brings me back to my customary way of living. ... I fall back of my own accord into my old opinions, and dread being awakened, lest the toilsome wakefulness which follows upon a peaceful rest must be spent thenceforward not in the light but among the inextricable shadows ... ([1641], p. 63)

For Hilbert, though, the toilsome wakefulness of skeptical foundationalism is not a challenge from which the promise of new levels of mathematical certitude awaits all who overcome their intellectual laziness. It is an unwelcome interruption of the mathematical dream that puts one in the contrived and unhelpful state of puzzlement and ineptitude where before all was in perfect order.

Thus is Hilbert's naturalistic epistemology. The security of a way of knowing is born out, not in its responsibility to first principles, but in its successful functioning. The successful functioning of a science, moreover, is determined by a variety of factors – freedom from contradiction is but one of them – including ease of use, range of application, elegance, and amount of information (or systemization of the world) thereby attainable. For Hilbert mathematics is the most completely secure of our sciences because of its unmatched success, and this unambiguous certainty is all the justification that any way of knowing should call for. If from some external perspective mathematics appears to be in jeopardy, this is evidence *against the tenability of that perspective*, not in favor of a skepticism about mathematics. Hilbert articulates this epistemic stance in a succinct summation of his analysis of Weyl's Predicativism:

Mathematicians have pursued to the uttermost the modes of inference that rest on the concept of sets of numbers, and not even the shadow of an inconsistency has appeared. If Weyl here sees an "inner instability of the foundations on which the empire is constructed," and if he worries about "the impending dissolution of the commonwealth of analysis," then he is seeing ghosts. Rather, despite the application of the boldest and most manifold combinations of the subtlest techniques, a *complete security of inference and a clear unanimity of results* reigns in analysis. We are therefore *justified* in assuming those axioms which are the basis of this security and agreement; *to dispute this justification would mean to take away in advance from all science the possibility of its functioning* ... ([1922], p. 200, emphasis added)

2.3 MATHEMATICAL AUTONOMY

If Hilbert recognized no *Grundlagenkrise* in mathematics, what, after all, was the point of his elaborate foundational program? Why endeavor so assiduously to demonstrate what one takes to be the unshakable starting point of all inquiry – the consistency of one's own methods?

For a certain type of naturalist, these questions may have no satisfactory answer, and so to answer them one must further sharpen one's understanding of Hilbert's epistemology.

The contrast here is with the anti-foundationalism of Wittgenstein's remarks in *On Certainty* and the naturalistic epistemology depicted there – the view according to which "at the foundation of well-founded belief lies belief that is not founded" (§253). Wittgenstein suggests that, for anyone, *some* system of belief must be completely immune from doubt because it is the system from which the person weighs the truth or falsity of claims, the ground on which he or she stands in order even to express doubt: "I have a world picture. Is it true or false? Above all it is the substratum of all my enquiring and asserting" (§162). "But I did not get my picture of the world by satisfying myself of its correctness; nor do I have it because I am satisfied of its correctness. No: it is the inherited background against which I distinguish between true and false" (§94).

If one could simply take mathematical methods to be constitutive of argumentative grounds, then it is clear both why the skepticism of certain foundational programs would not appear threatening and why doubt about mathematical methods would not in general arise. Should the skeptic point out that justification for certain principles, say mathematical induction, was lacking, one could only wonder at the question. Whatever justification there could be would have to be more certain than the principle itself in order to gain any ground, and that is unthinkable: Wittgenstein's remarks that "[his] not having been on the moon is as sure a thing ... as any grounds [he] could give for it" (§111) and that "[his] having two hands is, in normal circumstances, as certain as anything that [he] could produce in evidence for it" (§250) would apply just as well to mathematical induction.

Similarly, so long as one continues to work within the ordinary mathematical framework, the kind of doubt that is directed *at* that framework would be impossible:

All testing, all confirmation and disconfirmation of a hypothesis takes place already within a system. And this system is not a more or less arbitrary and doubtful point of departure for all our arguments: no, it belongs to the essence of what we call an argument. The system is not so much the point of departure, as the element in which arguments have their life. (§105)

Because it would not be couched within one's own argumentative standards, doubt directed at mathematical methods would be lifeless. Mathematical methods, then, just are never subject to serious doubt by virtue of their especial certitude.

To an extent this Wittgensteinian naturalism accords with Hilbert's position. Hilbert objects to the foundational programs of Brouwer and Weyl simply because they are rooted on non-mathematical grounds from which ordinary mathematics appears to be in need of justification. That is reason enough for Hilbert to object to those programs and reject the grounds on which they are rooted. Hilbert recognizes no rite of arbitration in any standpoint from which the legitimacy of the mathematician's methods falls into question. The legitimacy of those methods cannot reasonably be questioned.

But a thoroughgoing Wittgensteinian attitude about mathematics precludes any need for any foundational program. No epistemic gains are available if mathematics already is "the element in which arguments have their life." Yet Hilbert offers a foundational program and promises from it epistemological gains. Hence his position cannot be in full accordance with the Wittgensteinian's.

Distinctive of Hilbert's position is that mathematical methods are *neither* subject to scrutiny from any non-mathematical standpoint *nor* constitutive of our argumentative standards. Since mathematics is not subject to scrutiny, there is no foundational crisis to overcome. But the reason they are not subject to scrutiny is not the Wittgensteinian reason that such scrutiny would be lifeless or senseless by virtue of all meaningful or gainful scrutiny taking place already within the mathematical framework. For Hilbert not even mathematics plays the role of first principles. Certainly, at least, mathematics' high mark of certitude is not due to it playing this role.

This distinction between Hilbertian and Wittgensteinian naturalism is most evident in Hilbert's claim that mathematical methods *are* justified, in contrast with the Wittgensteinian principle that our epistemic bedrock is not and cannot be justified. According to Hilbert mathematics is justified, though not on any philosophical grounds: Mathematics is justified in application, through a history of successful achievement, through the naturalness with which its methods come to us, through its broad range of applicability, etc. This justification earns for mathematics a position of unassailability, but it does not earn for it the position of epistemic bedrock.

Indeed when Hilbert claims that the mathematical methods "which Aristotle taught and which men have used ever since they began to think" cannot be challenged because "no one ... could keep people from negating general statements, or from forming partial judgments, or from using *tertium non datur*" he says also that these methods "do not hold" in all the contexts in which mathematicians use them ([1926], p. 219). That the methods "do not hold" is precisely the skeptical challenge, and Hilbert's response again is extraordinary. He refutes the skeptic, not by disagreeing with the content of the challenge – that mathematical methods do not hold. He *agrees* with the skeptic and yet *still* declares the challenge inappropriate. The mathematician's duty is not to find laws that "hold" but ones that get justified in practice. Once they are so justified, not even the concession that they betray the most fundamental philosophical principles amounts to foundational crisis.

On this account, though, skeptical foundationalism still has a foothold. Hilbert feels that the justification of a measure arises solely from the measure's demonstrated success, and on that account his confidence in mathematics is as high as it could be. But unlike the Wittgensteinian naturalist who can perpetually check the skeptic, Hilbert has no recourse to the meaninglessness of skepticism about mathematics, he has only his particularly naturalistic grounds of justification against theirs. In the end, his position seems to him unshakable, because he can see how the skeptic's path leads to the death of all science and that mathematicians are unlikely to follow down it. But against the persistent skeptic this is not an argument of the sort that the Wittgensteinian could offer. The skeptic simply could call for the death of all science in deference to his philosophical scruples. Hence, even as it cannot mount a viable reformation, skepticism may endure. Hilbert would like the legitimacy of mathematics to speak for itself, but the skeptic has him appealing to certain standards of justification to defend his science.

This is the setting of Hilbert's program. Though the skeptic fails to make a case against the consistency and reliability of mathematics, his attack does enough damage if it exposes a dependence of the veracity of mathematical methodology on *any* justificatory standards. For even if questioning those standards is unreasonable, even if, that is, to do so "would mean to take away in advance from all science the possibility of its functioning," the status of mathematics is diminished if its veracity is shown to rest, through however circuitous a route, on non-mathematical grounds.

This explains Hilbert's at first puzzling approach to foundational research, his endeavor to prove what he emphasizes is in no way in doubt: the consistency of mathematics. For the epistemological gain to be earned with such a proof is not the knowledge that mathematics is consistent, it is the knowledge that mathematics need not appeal to anything non-mathematical in its own defense and that its truths are in that sense objective, "ultimate" truths:

Accordingly, a satisfactory conclusion to the research into these foundations can only ever be attained by the solution of the [mathematical] problem of the consistency of the axioms of analysis. If we can produce such a proof, then we can say that mathematical statements are in fact incontestable and ultimate truths – a piece of knowledge that (also because of its general philosophical character) is of the greatest significance for us. (Hilbert [1922], p. 202)

Thus it is evident why Hilbert takes his program to probe deeper than other foundational efforts. His adversaries' programs all treat as open questions whether and how much mathematics is consistent, and they aim to settle these questions by setting some portion of mathematics securely on some principles of first philosophy. By contrast, Hilbert *begins* his foundational research assuming that all orthodox mathematics is consistent and asks instead whether mathematics is autonomous in the sense that its consistency, and therefore its legitimacy, depends ultimately on *no* principles of first philosophy. Success would "regain for mathematics the old reputation or incontestable truth" by making its truths welcome to everyone *regardless of* the philosophical principles they endorse.

Since Hilbert's "consistency question" ultimately is the challenge of taking mathematics out of any philosophically informed setting, his estimation of others' attempts at solving the question is low:

The importance of our question about the consistency of the axioms is well recognized by philosophers, but in [the philosophical literature] I do not find anywhere a clear demand for the solution of the problem in the mathematical sense. ([1922], p. 201)

Without settling the problem mathematically, it is unclear what a "solution" to the problem even could gain, since, after all, Hilbert has no doubts about mathematics' consistency. Thus Bernays explains that

[t]he great advantage of Hilbert's procedure rests precisely on the fact that the problems and difficulties that present themselves in the grounding of mathematics are transformed from the epistemologico-philosophical domain into the domain of what is properly mathematical. ([1922b], p. 222)

This is such a great advantage, he says elsewhere, because "mathematics [thereby] takes over the role of that discipline which was earlier called *mathematical natural philosophy*" ([1931], p. 236). With mathematics itself in that role, Hilbert's defense can achieve a kind of unambiguity that mathematics deserves. Again Hilbert's words in his introduction to the Hamburg lectures are key: "in mathematical matters ...it should not be possible for half-truths or truths of fundamentally different sorts to exist" ([1922], p. 198). In particular it should not be possible to settle mathematical matters in ways that essentially favor any particular set of philosophical assumptions. Thus Hilbert's program embodies philosophical subtlety, for the gains he envisions are philosophical, but the program's realization depends on turning philosophical inquiry over to purely mathematical methods.

2.4 FORMALISM AND FINITISM

The ingenuity behind Hilbert's "formalism" and "finitism" lies in the role that these theses play in securing the transfer of philosophical inquiry into the mathematical domain. Each thesis amounts to a methodological guideline designed to ensure that the foundational program delivers the kind of unambiguous mathematical self-sufficiency described in the last section. That is, should the foundational program betray either principle, the program would fail, but Hilbert argues that if the consistency result can be proven in accordance with these principles, one will have shown that mathematics is beholden to no philosophical framework.

Thus it will not do to interpret "formalism" as the doctrine that mathematics is meaningless or that its subject matter consists just of formal symbols and rules of formula manipulation, as the term is often used in current philosophical discussions. Neither is it correct to understand Hilbert's "finitism" as the doctrine that only decidable methods are veracious and that only finitary propositions are contentful, as has been alleged in various forms since Kronecker. Hilbert explicitly wants to avoid appealing to any doctrines about the subject matter or ontology of mathematics. Indeed, as argued above, if research constrained by theses such as these proved inadequate to lay a foundation for all ordinary mathematics, then Hilbert would abandon those theses before yielding any of the mathematics. And if, alternatively, such a program succeeded, the resulting "defense" of mathematics would have the mathematical edifice resting on these metaphysical principles, which Hilbert hardly would consider an improvement over the defense already available in terms of mathematics' success in application. Both Hilbert's "formalism" and his "finitism," instead of being philosophical perspectives from which he intends to justify mathematical techniques, are methodological constraints *forced* by the type of mathematical self-reliance that he intends to demonstrate.

Following Mancosu ([1998a], p. 163)[2] let us note first that Hilbert nowhere describes himself or his outlook as "formalist." The label seems to originate instead in the polemic from representatives of other foundational schools intended to draw into question the legitimacy of "Hilbert's philosophical perspective." Aside from the notorious correspondence between Hilbert and Frege on the foundations of geometry, which in any case predates the foundational

---

[2] Mancosu attributes the label primarily to Brouwer [1928].

perspective that characterizes Hilbert's program,[3] the most often cited passage in support of attributing a "formalist" philosophy to Hilbert is the following: "The solid philosophical attitude that I think is required for the grounding of pure mathematics...is this: *In the beginning was the sign*" ([1922], p. 202). It will become clear, however, that even this proclamation must be understood as a description of the philosophical attitude that Hilbert feels one must adopt in order properly to engage in the foundational pursuit of mathematical autonomy, and not as a description of the correct theory concerning the nature or origin of mathematics.

Bernays describes the position explicitly in his reply to Aloys Müller's criticism of "Hilbert's conception of numbers as signs":

Hilbert's theory does not exclude the possibility of a philosophical attitude that conceives of the numbers as existing, nonsensical objects [as Müller would have them be]. ...Nevertheless the aim of Hilbert's theory is to make such an attitude dispensable for the foundations of the exact sciences. ([1923], p. 226)

Thus if formalism is supposed to be a type of nominalist or anti-realist metaphysical doctrine, then such cannot be consistent with this description of Hilbert's program. According to Bernays, success for Hilbert's program would not weigh in on the question of whether numbers exist, or whether alternatively mathematics consists solely in meaningless signs. Adopting the attitude that "in the beginning was the sign" serves rather to separate all answers to such questions from the foundational program.

---

[3] Detlefsen [1993] distinguishes developmental stages leading to Hilbert's invention of proof theory. Specifically he separates Hilbert's early remarks about the formal nature of axiomatics and the "hypothetical" role that axiomatic systems generally played from the more thoroughgoingly formalist perspective of Hilbert's foundational investigations in the 1920s according to which even the logical symbols are treated as meaningless.

Similarly Hilbert's call for a restriction to purely finitary or constructive techniques for the sake of foundational research is a strategy needed in order to secure mathematical self-reliance. Since Hilbert harbors no doubts about the reliability of any mathematical techniques, in a sense all of them are at his disposal. But on pain of circularity, some restriction is due for the purposes of evaluating and justifying the techniques themselves. In his own words:

> We therefore see that, if we wish to give a *rigorous* grounding of mathematics, we are not entitled to adopt as logically unproblematic the usual modes of inference that we find in analysis. Rather, our task is precisely to discover *why* ... we always obtain correct results from the application of transfinite modes of inference of the sort that occur in analysis and set theory. (Hilbert [1923], p. 1140, emphasis added)

Hence for programmatic purposes those same modes of inference that we seek to evaluate cannot figure into the evaluation, for in case they should, the original question as to why a result is correct could be put to the result *of the evaluation*.

To step out of this circle, any restriction of techniques would do. The resulting justification just will only ever be relative to the techniques that are required. One must begin somewhere, however, so the relativity of the evaluation *per se* is not a complaint against it. The foundational task, as Hilbert saw it, was to step back far enough that only techniques that everyone recognized as mathematically acceptable were used, while at the same time retaining resources sufficient to carry out the evaluation. From Hilbert's point of view, everyone's demands weigh in equally on this matter, because their several perspectives are constitutive of the skepticism that he wants to ward off. The appeal to finitary or constructive techniques, therefore, is not so much a recourse to foundations that Hilbert would argue were epistemologically secure, as a measure to ensure that all mathematics gets justified wholly within mathematics. Again, Bernays articulates the position:

One thus arrives at the attempt of a purely constructive development of arithmetic. And indeed the goal for mathematical thought is a very tempting one: Pure mathematics ought to construct its own edifice and not be dependent on the assumption of a certain system of things. ([1922b], p. 217)

For Hilbert in no way wants to abandon the constructive tendency that aims at the self-reliance of mathematics. ([1922b], p. 219)

Thus our development supports the interpretation of Howard Stein:

I think it is unfortunate that Hilbert, in his later foundational period, insisted on the formulation that ordinary mathematics is "meaningless" and that only finitary mathematics has "meaning." Hilbert certainly never abandoned the view that mathematics is an organon for the sciences: he states this view very strongly in the last paper reprinted in his *Gesammelte Abhandlugen*, called "Naturerkennen und Logik"; and he surely did not think that physics is meaningless, or its discourse a play with "blind" symbols. His point is, I think, this rather: that the mathematical *logos* has no responsibility to any imposed *standard* of meaning: not to Kantian or Brouwerian "intuition," not to finite or effective decidability, not to anyone's metaphysical standards for "ontology" ... ([1988], pp. 254–5)

The question remains as to *how*, according to Hilbert, the methodological constraints of formalism and finitism are forced on one by the demands of an earnest attempt at establishing mathematical autonomy. Answering this question brings out how under the naturalistic conception of Hilbert's program there is a lively interaction between the two principles.

The general method underlying Hilbert's program is familiar. One first fully formalizes a branch of mathematics as an axiomatic system so that one is dealing, not with mathematical statements and inferences, but with formulas and admissible sequences of formulas. Already one's subject matter has been "formalized," but the next step brings out the particularly "formalist" nature of the method: When one sets out to study this axiomatic system, and specifically when one undertakes to demonstrate its consistency, one must suspend throughout the investigation the original "meanings" of the statements that have been formalized. Bernays describes this suspension of interpretation as necessary:

> Accordingly, in Hilbert's theory we have to distinguish sharply between the formal image of the arithmetical statements and proofs as *object* of the theory, on the one hand, and the contentual thought about this formalism, as *content* of the theory, on the other hand. The formalization is done in such a way that formulas take the place of contentual mathematical statements, and a sequence of formulas, following each other according to certain rules, takes the place of an inference. And indeed no meaning is attached to the formulas; the formula does not count as the expression of a thought ... (Bernays [1922b], p. 219)

That is, the branch of mathematics one investigates becomes for the sake of the investigation a purely formal object. One "has to" proceed in this way, as Bernays says, in order that the branch of mathematics fit entirely under the lens of mathematical investigation. If, for example in one's demonstration that a branch of mathematics is consistent, one falls back to the pre-theoretical interpretation of formulas as statements, then the semantic assumptions behind that interpretation will have polluted the would-be purely mathematical achievement. In the Hamburg lectures, Hilbert describes this transition to proof theory in similarly normative terms: "To reach our goal, we *must* make the proofs as such the object of our investigation; we are thus *compelled* to a sort of 'proof theory' which studies operations with the proofs themselves" ([1922], p. 208, emphasis added).

As an example of the procedural guidelines that emerge from the formalist constraint, Bernays describes how the program's ultimate goal is shaped by it:

What in particular emerges from this consideration about the requirement and the purpose of the consistency proof is that this proof is only a matter of seeing the consistency of arithmetic theory in the literal sense of the word, that is, *the impossibility of its immanent refutation*. ([1931], p. 260)

This is in contrast to the at-the-time more familiar means of establishing consistency, which was to determine whether "the conditions formulated in the axioms can at all be satisfied by means of a system of objects with certain properties that are related to them" ([1931], p. 237). This route to the consistency of arithmetic is easily established through reference to the standard model of natural numbers. But the goal of mathematical autonomy demands a purely syntactic demonstration. For on the one hand the consistency proof by way of reference to the standard model rests on the semantic assumptions underpinning one's grasp of that model and its accordance with the arithmetic axioms. And on the other hand even if one's model theory were fully mathematized so that the demonstration of consistency in this way became rigorously mathematical, the amount of mathematics involved would of course far extend the arithmetic theory under investigation, resulting again in justificatory circularity.

Meanwhile, since if the evaluation is to be genuinely mathematical then some mathematics must be assumed through the course of one's proof-theoretical investigations, this base mathematics need not, and in fact cannot, be stripped of its meaning. One must use it and work within it in order to reason about the formal axiomatization that one is studying, thereby attaining results relative to the reliability of that base mathematics:

[I]n addition to this proper mathematics, there appears a mathematics that is to some extent new, a *metamathematics* which serves to safeguard it by protecting it from the terror of unnecessary prohibitions as well as from the difficulty of paradoxes. In this metamathematics – in contrast to the purely formal modes of inference in mathematics proper – we apply contentual inference; in particular, to the proof of the consistency of the axioms. (Hilbert [1922], p. 212)

Hence Hilbert's famous two-tiered approach to foundational studies.[4] One distinguishes the formal object of investigation and the contentual base in which the investigation is carried out:

In this way the contentual thoughts (which of course we can never wholly do without or eliminate) are removed elsewhere – to a higher plane, as it were; and at the same time it becomes possible to draw a systematic distinction in mathematics between formulae and formal proofs on the one hand, and the contentual ideas on the other. ([1922], p. 204)

The formalist nature of Hilbert's *Beweistheorie* therefore arises from the need to eliminate philosophical assumptions from one's meta-mathematical investigations. With the semantic assumptions stripped away, what remains for one's scrutiny are only "signs," uninterpreted formulas. The remaining question is where to delineate the contentual base of meta-mathematics, that bit of mathematics that is spared strict formalization. One point of consideration is that the contentual base be weaker than the "proper mathematics" in order that the justification not exhibit circularity. Another is that, ideally, this base should be weak enough not to be the target of the skepticism from the rival foundationalist schools. In addition to these, a third constraint now arises, which is that the contentual

---

[4] Notice that Wittgenstein often seems to describe such an approach as ruled out from the start in remarks such as the following from his *Remarks on the Foundations of Mathematics*: "What I want to say is this: mathematics as such is always measure, not thing measured" (III-§75). I discuss Wittgenstein's and Hilbert's differing interpretations of meta-mathematics in Section 6.3.

base theory should be *strong* enough to allow one to reason within it effectively about signs and sequences of signs. At first it is not evident whether all three conditions can be met. That is, there is a question whether any mathematics could exhibit the deductive strength needed for robust investigation of formulas as such without extending in strength the arithmetic theory Hilbert intends to defend. And even if this circularity is avoidable, the degree of achievement is only partial if the meta-mathematics needed still is strong enough to be in need of its own defense. Hilbert claims that something called finitary mathematics meets all three criteria.

Bernays gives the clearest statement of how the contentual base theory is determined by the demands of formalism:

Now the only question still remaining concerns the means by which this proof should be carried out. In principle this question is already decided. For our whole problem originates from the demand of taking only the concretely intuitive as a basis for mathematical considerations. Thus the matter is simply to realize which tools are at our disposal in the context of the concrete-intuitive mode of reflection. ([1922b], p. 221)

Exactly what Bernays means by "concretely intuitive" is the subject of considerable debate, both in terms of the philosophical nature of this mode of reflection[5] and in terms of the answer, in the mathematical sense, to his question as to which tools are available in this reflection.[6] When he speaks of "the demand of taking

---

[5] I.e., whether this mode corresponds with an empirical faculty or, as Hilbert and Bernays suggest in some later writings, with a Kantian intermediate faculty between experience and thought.

[6] Tait [1981] argues that Hilbert's finitary mathematics is Primitive Recursive Arithmetic. Others, for example Volker Halbach in private conversation, have argued that it is hard to imagine Hilbert rejecting as foundationally significant Gentzen's arithmetic consistency proof, had it been available, on grounds that it was not finitary. Evidence for the latter interpretation is found in the Introduction of the second edition of Hilbert and Bernays's *Grundlagen der Mathematik*, where Bernays discusses extending the finitist standpoint in order to tackle new questions. Were the bounds of finitism derived from philosophical

only the concretely intuitive as a basis for mathematical considerations," however, Bernays can only mean the demand imposed
on foundational studies by the fact that these studies must proceed
uninfluenced by any philosophical considerations. That is, Bernays
is referring to the demand of investigating proper mathematics
purely formally. Hence the proper delineation of meta-mathematics
is to be determined by isolating that minimal fragment of proper
mathematics sufficient to investigate purely syntactic aspects of formal, axiomatic theories, and not by unpacking the exact nature of
"concretely intuitive" reflection in the philosophical sense.

On the other hand, Hilbert does want to say something about
the philosophical nature of finitary mathematics, specifically that
it is unassailable on skeptical grounds. Thus he argues that not
only is this amount of mathematics necessary for metatheoretical
evaluation because of the requirement that one be able to reason
effectively about formulas, but that also it is sufficiently minimal
to be beyond criticism in something more like the Wittgensteinian
sense:

If logical inference is to be certain, then these objects must be capable of
being completely surveyed in all their parts, and their presentation, their
difference, their succession (like the objects themselves) must exist for us
immediately, intuitively, as something that cannot be reduced to something
else. (Hilbert [1922], p. 202)

That is, since the subject matter of meta-mathematics is purely formal, it is fully concrete, finite, survey-able, and immediate. Thus
meta-mathematical reasoning need *only* deal with the recognition
of and distinction between concrete, immediately present objects
and need *not* recapture any interpretation of these objects. Hilbert

insight rather than details of practice, such an extension would not likely be possible.
Zach [1998] develops this idea.

claims that this is a kind of bedrock of reasoning, irreducible and consequently unchallengeable.

Hilbert's claim that there need be no defense for finitary mathematics is of course controversial – the majority of the controversy due to his unclarity with respect to what exactly finitary mathematics amounted to. A central example occurs with the reasoning involved in a demonstration of consistency: There is something characteristically finitary about the verification that a single proof involves no contradiction as well as the verification that if any proof of a particular form is free of contradiction then so is another attainable from it through a constructive transformation. But one must use some principle of induction in order to reason from these points to the consistency claim that no proof contains a contradiction. And this principle of induction is not obviously finitary in any philosophically uncontroversial sense. This very point was the crux of the debate between Oskar Becker and Hilbert with respect to the nature of induction admissible in meta-mathematical reasoning.[7]

What should strike one as even more controversial about Hilbert's claim, however, given his steadfast commitment to a naturalistic view of mathematics, is the very fact that he wants at this point to appeal to the epistemic status of finitary reasoning. He cannot, after all, simply be hoping to secure mathematics on a foundation of the "concrete intuitive mode of reflection." Despite all his remarks in favor of the solidity of finitary reasoning, such a conception would just reduce his program to a fraudulent attempt at *ignotum per ignotius* – explaining what is unknown by what is more

---

[7] See Mancosu [1998a], pp. 165–7. Importantly, Becker focuses his criticism on the need for Hilbert to move beyond the purely finite in his meta-mathematics, while Hilbert replies only that meta-mathematical induction is different and weaker than the "full induction axiom" of proper mathematics.

unknown – and Hilbert's principal aim is to avoid precisely this sort of foundationalism.

Rather, Hilbert's appeal to the especially basic status of the "concrete intuitive" is better understood as a strategic advertisement for his program. He is confident that he has found a way to provide a purely mathematical evaluation of mathematics itself, the essential device in doing so being the formalist perspective in proof-theoretical investigations. And he sees that the evaluation can proceed thus in a philosophically gainful way, since the meta-mathematics can thereby be constrained to principles weaker than those comprising the formalized theory of mathematics proper. Already, then, the program amounts to a significant achievement. But this accomplishment might be lost on the broader philosophical and mathematical communities if the techniques involved cannot be shown to be everywhere sound, not by Hilbert's standards but by theirs. It is therefore something of a happy accident if in fact the meta-mathematics falls within a "finitary" rubric acceptable even to the most skeptical Predicativists, Intuitionists, and other mathematical cautionaries. If it does not, then still the degree of self-sufficiency that is attainable makes headway at demonstrating the needlessness of philosophical grounding for mathematics.

So just as Hilbert's formalism is a procedural consequence of the demands of his unique epistemological goals, his finitism is necessitated by that formalism. That is, so long as one's meta-mathematical evaluation steers away from any principles other than those needed to reason directly about the strictly formalized axiomatization of ordinary mathematics, then one is on track to uncover a purely mathematical appraisal of mathematics itself. Whether or not the "finitism" inherent in that course can truly be said to be finitary in every philosophically informed understanding of the term, and whether or not as a consequence it seems as epistemically innocuous as Hilbert describes it as being are at most secondary considerations.

2.5 CONCLUSION

In another report from 1922 entitled "Hilbert's significance for the philosophy of mathematics" where he focuses specifically on the innovation of rigid axiomatization, Bernays discusses the nature of Hilbert's earlier achievements in the foundations of geometry:

[A] new sort of mathematical speculation [had] opened up by means of which one could consider the geometrical axioms from a higher standpoint. It immediately became apparent, however, that this mode of consideration had nothing to do with the question of the epistemic character of the axioms, which had, after all, formerly been considered as the only significant feature of the axiomatic method. Accordingly, the necessity of a clear separation between the mathematical and the epistemological problems of axiomatics ensued. ([1922a], pp. 191–2)

We have seen that a very similar separation between mathematics and epistemology characterizes the foundational innovations that Hilbert was introducing that very year in his pursuit of arithmetical consistency, specifically by deepening foundational research "not so much to fortify individual mathematical theories as because, in my opinion, all previous investigations into the foundations of mathematics fail to show us a way of formulating the questions concerning foundations so that an unambiguous answer must result" ([1922], p. 198).

In the next sentence Bernays traces the origin of this mathematical, as opposed to epistemological, attitude in foundational studies to Felix Klein: "The demand for such a separation of the problems had already been stated with full rigor by Klein in his Erlangen

Programme." Here is Klein's [1908] description of his conception of the nature of foundational research:

Mathematics has grown like a tree, which does not start at its tiniest rootlets and grow merely upward, but rather sends its roots deeper and deeper at the same time and rate as its branches and leaves are spreading upwards. Just so – if we may drop the figure of speech – mathematics began its development from a certain standpoint corresponding to normal human understanding and has progressed, from that point, according to the demands of science itself and of the then prevailing interests, now in the one direction toward new knowledge, now in the other through the study of fundamental principles.

It would be incorrect to infer from this image, however, that Hilbert's foundational pursuits were not philosophically motivated, that his naturalistic conception of the formal sciences amounted simply to a disinterested rejection of epistemological concerns about mathematics. Bernays' point, rather, is this: that Hilbert's efforts in axiomatization and studies of fundamental principles are not efforts directed at uncovering epistemic foundations *in the axioms*; the legitimacy of an axiomatization is earned purely mathematically, through its ability to realize mathematically prescribed goals, and not by way of the *epistemic character* of the axioms. However, the mathematical goals put to any particular foundational program can very well arise from epistemological concerns. Hilbert's own was the desire to demonstrate that mathematics was a fully self-supporting science, and that mathematical certainty therefore enjoyed an epistemic status privileged beyond that of any way of knowing that rested on a philosophical conception of justificatory grounds.

CHAPTER 3

Arithmetization

3.1 INTRODUCTION

Hilbert's program laid the groundwork for a demonstration of the autonomy of mathematics. We have stressed that this set-up itself is of philosophical significance, for the approach Hilbert articulated uncovers a precise response to long-standing epistemological questions about justification. Precisely, Hilbert's innovative stance was this: Mathematics is a way of knowing that cannot and need not be justified on any *a priori* grounds. As such, it is not properly the target of skeptical attacks, which in essence demand such grounds. Nevertheless mathematics can be the subject of foundational insight, through a self-evaluation the outcome of which is that questions about how and why mathematical techniques work the way they do can be given purely mathematical answers. Thus in his numerous remarks in the vein of the following passage, Hilbert speaks of a type of "security" that cuts deeper than mere claims to knowledge:

Just as the physicist investigates his apparatus and the astronomer investigates his location; just as the philosopher practices the critique of reason; so, in my opinion, the mathematician has to secure his theorems by a critique of his proofs, and for this he needs proof theory. (Hilbert [1922], p. 208)

The "security" to be established by proof-theoretical critique is not assurance that mathematical techniques are reliable. It is a second-level claim about the nature of their reliability. Hilbert wants to demonstrate that their reliability is self-witnessing and not founded on any non-mathematical base.

In his polemical writings and speeches Hilbert often expresses distaste with challenges drawn against the legitimacy of mathematics, not because of the content of the charges advanced but because of the inherently subjective armory behind them. Thus Brouwer's rhetorical edge, we saw in the last chapter, was that he spoke "with the tongue of angels." But even the best rhetoric is still just rhetoric. For his defense Hilbert does not take on his critics directly. Instead he suggests we take a second look at mathematics and realize that something unique about it makes it immune to their attack. Thus we are not to be swayed "just because of Kronecker's pretty eyes" or "just because a few philosophers ... have put forward reasons" however enchanting or compelling these might be ([1931], p. 268). Once we see that mathematics does not rest on anything but itself, no challenge to its veracity or use stemming from outside mathematics will seem relevant. The trail of philosophy never crosses into mathematical terrain.

This distinguishes Hilbert's epistemological views from similar anti-foundationalist programs according to which such self-evaluation would be viciously circular or, at best, redundant. As a consequence, Hilbert tells us, mathematics enjoys a kind of objectivity, distinguishing it both from sciences that require grounding in particular philosophical systems as well as from those whose principles are not open to evaluation.

The leading insight behind Hilbert's program is that the proper reasoning about mathematics is itself mathematical, and that if one studies mathematics purely formally, then the mathematics needed *for* this study is by design free of semantic assumptions and significantly weaker than the mathematics that one is therewith able *to* study. Thus one is able, in a single step, to assure oneself that the ensuing evaluation is independent of any commitment to specific *a priori* or philosophical grounds and also to sidestep circularity.

Of course, in order for this approach to succeed, one's proof theory must be properly mathematical. Hilbert's vision for his program and what it was designed to accomplished is a significant contribution to philosophical thought since therein lies his unique perspective on the foundations of mathematics and devastating criticism of traditional philosophical foundationalism. But this vision cannot stand on its own since, according to Hilbert, the thesis that mathematics is autonomous hangs on the prospects of a rigorously mathematical science of metatheory. That is, one could be committed with Hilbert not to turn questions about the legitimacy of mathematical methods over to philosophical speculation regardless of the success or failure of his program, but without its success one could not expect others to share this commitment. Hilbert did not wish to argue his adversaries out of their convictions and into his own. Indeed he could not do this. Any attempt would be self-defeating because he held the foundations of mathematics to be beyond argumentation. His strategy was instead to show from within mathematics why there is never reason to step outside mathematics to determine why and whether its methods work.

In this crucial sense, though, Hilbert's preliminary results were fully inadequate. If one considers the consistency proofs he sketched for fragments of arithmetic, one encounters considerable technical vagueness. In particular it is not easy to pin down exactly how much computational strength is needed to carry out the proofs, and – what is even worse – one finds no hint how even to formulate strictly mathematically the claim that an arithmetical system is consistent. A traditional skeptic wanting to re-situate mathematics on a new foundation of relatively secure philosophical principles might even have cited the partial results, confusions, errors, and ambiguities in the Hilbert school as reason for this impulse. Hilbert's program needed to be carried further.

In this chapter we look at techniques for the arithmetization of metatheory as the essential methodological contribution to Hilbert's program. It is through arithmetization that questions about a system of mathematics and its methods can be put to that system on its own terms, so that mathematics can "speak for itself." Two distinct approaches to arithmetization have been taken up since the 1920s, so my primary concern will be to distinguish these and to determine the nature of the contribution each makes to Hilbert's program. We will then be in a position to determine what answers mathematical systems return when one puts to them questions about themselves.

3.2 GÖDEL'S WORK AS A CONTRIBUTION TO HILBERT'S PROGRAM

Gödel's [1931] invention of a technique for the arithmetization of syntax is perhaps the most significant positive contribution to Hilbert's program. For by way of arithmetization, proof theorists can construct a sentence in the language of arithmetic the truth conditions for which are precisely those of an informal claim of that system's consistency. The question of the consistency of a formalized mathematical theory T can then be put to an arithmetical system in the form of the provability or refutability of an arithmetical sentence. As a consequence one can investigate which arithmetic theories "prove T's consistency" and which do not. Hence one can carry out the evaluation of mathematics in far more unambiguously mathematical terms. Our aim is to determine just how unambiguously mathematical those terms are. Prior to that investigation, however, let me briefly acknowledge that Gödel's work usually is thought of as a refutation of Hilbert's program and defend the contrary suggestion that some good news can be gleaned from it.

Since Gödel also proved that almost no consistent axiomatization of an appreciable amount of mathematics can in this way "prove its *own* consistency," the notion that his invention of arithmetization is a *positive* contribution to Hilbert's program has been de-emphasized. Indeed most commentators, reasoning as follows, have not construed it as a positive contribution at all: Gödel's second theorem reveals that the clarificatory gains to be found in the arithmetization of syntax are exactly the tools needed to show that the only way to prove the consistency of a mathematical system is to use mathematical techniques that extend those of the system itself. Therefore any mathematical defense of mathematics is circular in the sense that Hilbert hoped to show it need not be – hardly good news to a Hilbertian.

However, we have seen that the consistency proofs Hilbert sought were not his program's sole goals. Hilbert did not doubt the consistency of any branch of mathematics and did not see his program as testing, in any traditional sense, for consistency. Thus he would not conclude from a system's inability to prove its own consistency that its consistency was dubious. It is clear that Hilbert would have welcomed a non-circular, purely mathematical demonstration of the consistency of analysis and even that he expected such a result. This followed from his not being in a position in terms of practical expectation to distinguish the availability of such a demonstration from a different goal, the purely mathematical *formulation* of the statement that a branch of mathematics was consistent. Yet it is this latter goal that Hilbert emphasized when stressing the philosophical point of his program – that of formulating metatheoretical questions in such a way that unambiguous answers to them would result. Only because he was unfamiliar with incompleteness phenomena did Hilbert conclude that if such a formulation were available, then given the ample reason from scientific practice for believing in the consistency of ordinary mathematics the answer to the consistency question thus formulated would be an unambiguous "yes." But if Gödel's discovery of the arithmetizability of syntax dispelled Hilbert's optimism by forcing a distinction between the ability to put to a mathematical system the question of its own consistency and the system's ability to settle that question in the affirmative, it is only because the same discovery vindicated Hilbert's principal philosophical conviction: that by being cast within mathematics itself important questions *about* mathematics could be investigated without favoring any philosophical tendencies over others.

Strictly speaking then, something like Gödel's achievement is just what Hilbert's program needed to get off the ground properly. If Gödel's methods turn out to be the best way to execute a Hilbertian investigation of metatheory, then it is unfortunate that the non-circular consistency proofs Hilbert was after are not available through them. This tempers considerably Hilbert's optimism for the results his program would usher in, but it does not show that his unique perspective on mathematical foundations is untenable: One still has every reason to believe in the consistency of ordinary mathematics, and one has shown how to formulate the consistency question within mathematics. Hence now, if the skeptic would have the mathematician doubt his or her methods by exhibiting a vicious circle or the like, then he or she is invited to try to translate the antinomy into a language the mathematician is ready to listen to, into purely arithmetic terms in fact. In this way the mathematical naturalist can quiet the skeptic. For even if the skeptical challenge cannot be mathematically refuted, the mathematical formulation of metatheoretical questions at least puts the burden of proof on the skeptic.

A much worse situation would result, however, if nothing like Gödel's techniques were available, so that questions of consistency, completeness, and the like could only ever be thoroughly investigated outside the arena of mathematics. Gödel saved Hilbert's program from that kind of failure by describing the most crucial step out of the problem that plagued Hilbert's program from the beginning, the problem of formulation.

Finally it is interesting to note that Gödel's major, original development in his [1931] paper was just this good news for Hilbert's program. His two incompleteness theorems follow without much work from an application of the fixed point theorem of Rudolf Carnap [1934] and the traditional analysis of paradoxical sentences like the "liar sentence." Thus the substantial analytical work preceded him. The innovation in Gödel's paper was the technique he used to construct a sentence that was equivalent in arithmetic to the statement that it itself is unprovable. The technique is of course the arithmetization of syntax by way of "Gödel numbering" and is what allowed Gödel to apply the fixed point theorem and the analysis of paradoxical statements to the case of provability. (It is possible to see the seeds of this idea already in Hilbert's own writing. On page 199 of [1926] and page 471 of [1928] Hilbert wrote: "But a formalized proof, like a numeral, is a concrete and surveyable object.") Thus the primary contribution to Hilbert's program that Gödel made with the publication of his incompleteness theorems really is the positive one discussed here, from which the bad news may properly be seen as a corollary result.

I turn now to the history of the development of arithmetization and an analysis of its significance *vis-à-vis* mathematical autonomy.

3.3   THE *GRUNDLAGEN DER GEOMETRIE*

"Arithmetization," generally conceived, is any process by which
essentially non-arithmetical subject matter is "reduced" to the
theory of positive integers. Under this very general conception,
Dedekind's and Kronecker's foundational works were pioneering
examples of arithmetization – Dedekind through his construction of
the real numbers from sequences of rational numbers, which themselves were constructable directly from the integers, and Kronecker
through investigating how core areas of many branches of mathematics could be recovered through constructions of decidable sets
of integers.

Hilbert's early foundational work, too, exhibited arithmetization techniques, particularly in his [1899] work on the *Foundations
of Geometry*. Hilbert's project there was to show that Euclidean
geometry was axiomatizable in a finite number of logically independent axioms and that these axioms were complete and consistent.
The problems of logical independence and consistency in particular
led Hilbert to the study of interpretation through model construction. This was a significant foundational breakthrough on two
counts. First, the technique proved successful. Hilbert was able to
demonstrate through model constructions that a set of axioms for
Euclidean geometry was both logically independent and consistent,
while prior to his work it was not clear how to answer rigorously
either question. The second point of significance is that the success
of this technique convinced Hilbert of the value of the purely formal approach to axiomatization. In different guises this commitment
to formality would characterize mathematically and philosophically
his several foundational programs for decades.

It was the model constructions in *Foundations of Geometry* that
involved a type of arithmetization. Hilbert constructed models

specifically out of the positive integers, interpreting the geometric relations in Euclidean geometry as relations on a subset of the algebraic field so that, under this interpretation, the axioms of Euclidean geometry expressed arithmetical truths.[1] Thus the ability to derive a contradiction from the Euclidean axioms translates as a proof of contradiction in arithmetic.[2] Hilbert thereby reduced the consistency of Euclidean geometry to the consistency of arithmetic.

For many of Hilbert's contemporaries, this kind of arithmetization carried a special epistemological importance. For if the consistency of geometry can be shown to depend only on the consistency of arithmetic, then owing to the foundational role of the natural numbers and the intuitive truth in the axiomatization of operations on them geometry itself attains grounding in arithmetical intuition, which is to say, certainty.

Naturally enough this was not Hilbert's appraisal of the situation. The "arithmetization" of geometry was not for him a way to ground geometric truth in arithmetic truth, but rather a methodological step essential for a mathematical demonstration of the consistency and mutual independence of the Euclidean axioms. Nowhere in *Foundations of Geometry* does he applaud the epistemological status of arithmetical truth or claim that the consistency of arithmetic is self-evident or less questionable than that of geometry. His discussion is more modest. In his concluding remarks he writes:

In this investigation we have taken as a guide the following fundamental principle; *viz.*, to make the discussion of each question of such a character as to examine at the same time whether or not it is possible to answer

---

[1] Precisely, Hilbert's model Ω had a finite domain generated by application of the operations addition, subtraction, multiplication, division, and $\lambda x.\sqrt{1 + x^2}$.

[2] He concludes his demonstration with this sentence: "From these considerations it follows that every contradiction resulting from our system of [geometric] axioms must also appear in the arithmetic related to the domain Ω" (p. 29).

this question by following out a previously determined method and by employing certain limited means. (pp. 130–1)

Hilbert tells us that this principle is designed in order to lend "purity" to one's methods. Rather than being a route to epistemological bedrock, purity here refers to the degree to which one's methods can be said to be mathematical. When determining to which means one should limit oneself, one therefore must attend to the specific problem one wants to solve and ask which methods would turn this problem into a purely mathematical one:

> This fundamental principle, which we ought to bear in mind when we come to discuss the principles underlying the impossibility of demonstrations,[3] is intimately connected with the condition for the "purity" [*Reinheit*] of methods in demonstration, which in recent times has been considered of highest importance by many mathematicians. ... In fact, the preceding geometrical study attempts, in general, to explain what are the axioms, hypotheses, or means, necessary to the demonstration of the truths of elementary geometry, and it only remains now for us to judge *from the point of view in which we place ourselves* as to what are the methods of demonstration which we should prefer. (pp. 131–2, emphasis added)

Since one will find oneself operating from a different point of view depending on the system one wants to prove consistent, the methods one will choose with an eye set at mathematical purity also will differ from system to system. Hilbert's choice of an arithmetical interpretation of the Euclidean axioms thus has nothing to do with the epistemological status of arithmetic. It is the appropriate choice because the relative consistency proof that results is unambiguously mathematical. Indeed, Hilbert explains that a consistency proof relative to the theory of real numbers would also have been

---

[3] Independence and consistency proofs both are proofs that certain proofs are impossible. For independence proofs one proves that a certain axiom is not derivable from a set of other axioms; consistency proofs are proofs that the negation of one from a set of axioms is independent from that set.

possible. His preference for the algebraic domain Ω is not due to the consistency of the theory of reals being more dubious, but rather to the algebraic domain being countable, hence simpler, yet still sufficient for the mathematical task at hand (pp. 29–30).

This dependence of metatheoretical methods on the theory one investigates brings out another aspect of Hilbert's naturalistic approach to foundational studies. If the purpose of foundations is not to vouchsafe questionable mathematics by reducing it to some privileged theory – if indeed there is no privileged theory – then one would expect Hilbert to pursue foundations of branches of mathematics wherever possible, even when the most skeptical researchers are satisfied without foundations. Whether philosophers will tend towards skepticism about a system's consistency has no bearing on whether that system's consistency can be investigated as a mathematical problem. That Hilbert envisions his research this broadly is evident from a comment he makes about Kronecker's foundational perspective:

Kronecker, as is well known, saw in the notion of integer the real foundation of arithmetic;[4] he came up with the idea that the integer – and, in fact, the integer as a general notion (parameter value) – is directly and immediately given; this prevented him from recognizing that the notion of integer must and can have a foundation. I would call him a dogmatist, to the extent that he accepts the integer with its general properties as a dogma and does not look further back. ([1904], p. 130)

This ironic appraisal of Kronecker is indicative of Hilbert's naturalistic epistemology. Today one expects to hear Kronecker referred to as a dogmatist in the *negative* sense that he dogmatically withheld belief from all but finitary mathematics. Yet Hilbert refers here to Kronecker as dogmatic in a *positive* sense, for failing to recognize

---

[4] Hilbert here refers to arithmetic quite generally – what today one calls "analysis."

a need or possibility for foundations of the concept of the integers. But why look further back than Kronecker? If Hilbert will not grant even the integers as an epistemologically secure foundation, one wonders, what will he grant? The answer is: nothing. For to Hilbert mathematical foundations are to provide, not epistemological security, but an analysis of mathematical techniques. Such a provision is to be pursued anywhere that analysis can be illuminating.

Hence Hilbert did not view his own work in the *Foundations of Geometry* as reducing the epistemological status of geometry to arithmetic.[5] The value of his consistency proof, rather, was simply that it showed for the first time that the consistency of geometry could be proved mathematically and was therefore not dependent on grounding in Kantian intuition and the like. Hilbert often complained that philosophical musings about the nature of mathematics "blocked the path to analysis." In the last chapter we saw Hilbert use this phrase to mean that if one is too eager to view mathematics in a particular way – say as resting on certain ontological principles or modes of cognition – then one risks overlooking or declaring as illegitimate a great deal of mathematics that does not fit that picture. Importantly, though, he means also that once one takes oneself as having uncovered the "real foundation" of mathematics, one is prone to overlook the possibility of a mathematical evaluation of the part of mathematics that under that conception has an elementary status. An evaluation of it would serve no further foundational goal. Hilbert criticizes Poincaré thus in his Hamburg lectures:

Poincaré was from the start convinced of the impossibility of a proof of the consistency of the axioms of arithmetic. According to him, the principle of complete induction is a property of our mind ... His objection that this

---

[5] It is worth noting that Kronecker did not either. For him, all of mathematics should be arithmetized in some sense or another *except* for geometry, which had its own epistemological character.

principle could never be proved except by the use of complete induction itself is unjustified and will be refuted by my theory. ([1922], p. 201)

According to Hilbert, Poincaré's conception of induction as a property of our minds "blocks the path to analysis" by leading him to believe that the soundness of induction could never be proved. But a philosophy of mathematics that leads one to think of certain types of analysis as unnecessary (Why prove a property of our minds that is itself already at the heart of mathematical reasoning?), or as impossible (Will one not inevitably call upon that very principle one is trying to prove in the course of the proof?) is a handicap. Just like ordinary mathematics, foundational research must always, whenever mathematically possible, be pushed further.

3.4 HILBERT'S PROTO-PROOF THEORY

Bernays explains that Hilbert's interest in a consistency proof for arithmetic arises partly because of his disinterest in philosophical foundations: "One will recall that the consistency of Euclidean geometry was already proved by Hilbert by the method of reduction to arithmetic. Thus it now seems appropriate to prove the consistency of arithmetic by reduction to logic" ([1922b], p. 216). That is, Hilbert's attention turned to the problem of arithmetical consistency largely for the mundane reason that it was "the next" obvious place to push his research. Hilbert did not wish to rest content with the "epistemological security" found in the reduction of mathematics to arithmetic, as we have seen both because he was not interested in that kind of security and because not everyone agreed that arithmetic was secure. The problem, as he saw it, was to mathematically prove the consistency of all mathematical theories, including arithmetic, without recourse to philosophical principles. This would serve the double purpose, first of warding off skepticism by showing how to formulate questions about mathematical techniques within mathematics, and second of solidifying proof theory as a viable branch of mathematics research. The challenge that he faced was that Poincaré's suspicion about the impossibility of a non-circular proof of the consistency of arithmetic – one that did not utilize the full principle of induction – turned out to be not so easily refuted.

In [1922b] Bernays formulated three questions that he takes to characterize the structure of Hilbert's proof theory:

1. The constructive development must represent the formal image of the system of arithmetic and at the same time must constitute the object for the intuitive theory of consistency. How does such a development take shape?
2. How is the consistency statement to be formulated?
3. What are the means of contentual consideration through which the consistency proof is to be carried out? (p. 220)

Of course, the three questions are closely related. The first question is just about how to construct the object theory that one will investigate. Since Hilbert explicitly seeks to avoid any considerations of the meanings of formulas at this level, the construction must be purely formal. Of particular importance is the question of what counts as a proof. Hilbert's idea is that an object level *proof* is a finite sequence of formulas, each of which is either an axiom, a substitution instance of a formula that occurs earlier in the sequence, or the result of an application of *modus ponens* to two formulas that occur earlier in the sequence. The formula appearing at the end of the proof-sequence is the proved formula.

One of Hilbert's logico-arithmetic axioms is the formula

$$a = b \rightarrow (a \neq b \rightarrow \bot)$$

containing a logical primitive symbol for contradiction. The answer to the first question, therefore, suggests an obvious answer to the second one: "The statement of consistency is now simply formulated as follows: ⊥ cannot be obtained as the end formula of a proof" (p. 221). Simple as this is, Hilbert here departs from the foundational techniques of [1899]. In the earlier work, Hilbert "proved" consistency through interpretation, that is relative to the consistency of arithmetic, whereas now for the proof of the consistency of arithmetic he explicitly avoids all interpretation. This is because at this stage of the foundational project no semantic assumptions can be made. Arithmetic in a sense carries the burden of all of mathematics. Since the consistency of other branches can be demonstrated through arithmetical interpretations of their axioms, a semantics-free grounding of arithmetic would be a demonstration of the autonomy of all these branches.

As emphasized in the previous chapter, the answer to the third question is a direct consequence of the answers to the other two: The means through which the consistency proof is to be carried out are just whatever means are required for reasoning effectively about formal systems. This is welcome news on the one hand, for evidently whatever is required for this is adequately free of semantic assumptions:

Certainly the basic possibility and feasibility of the modes of reflection demanded [in reasoning about formal systems] can already be recognized by what has been said so far; and one sees that the considerations to be employed here are *mathematical* in a very general sense. (p. 221)

However, it is here also that Poincaré's circularity worry is vindicated. For when Bernays begins to enumerate the techniques required for the inspection of formal proofs, he lists induction:

This much is certain: We are justified in using the elementary ideas of sequence and ordering, as well as the usual counting, to the full extent. (For example, we can determine whether there are three occurrences of the sign → in a formula or fewer.)

However, we cannot get by in this way alone; rather, it is absolutely necessary to apply certain forms of complete induction. Yet, by doing so we still do not go beyond the domain of the concrete-intuitive. (p. 221)

By not going beyond the domain of the concrete-intuitive, Bernays means that in using some form of "complete induction" in one's proof theory, one does not thereby undermine the program by smuggling in "considerations" that are not "mathematical in a very general sense." Supposing that Bernays is right about this does not put to rest worries about a vicious circularity in the program, though. For if the proof-theorist must use forms of complete induction in order to demonstrate the consistency of arithmetic, it is questionable whether the resulting proof even is informative. Since the strength of arithmetic lies essentially in the principle of induction, assuming only a weak background theory from which the consistency of arithmetic is an open question presumably involves suspending the use of this principle. Thus if the machinery needed for a formal treatment of proofs already includes the principle of "complete induction," then Hilbert has hoist himself with his own petard, confirming Poincaré's claim in his attempt to refute it.

Hilbert naturally was sensitive to this concern, given his announcement that with his proof theory he would refute Poincaré's charge. So it is not surprising that on his behalf Bernays offers a way out of the circularity:

In this regard, two types of complete induction are to be distinguished: the narrow form of induction, which relates only to something completely and concretely given, and the wider form of induction, which uses either the general concept of whole number or the operating with variables in an essential manner.

Whereas the wider form of complete induction is a higher form of inference whose justification constitutes one of the tasks of Hilbert's theory, the narrower form of inference belongs to the primitive intuitive mode of cognition and can therefore be applied as a tool of contentual inference. (p. 221)

Circularity is to be avoided by using only this "narrow form of induction" in one's meta-mathematical evaluation, securing with it the more general "wide form." This suffices perhaps to avoid a kind of ontological circularity. Bernays seems to be explaining, that is, that one need only assume that induction is permissible when reasoning about formal signs. From this assumption one will be able to conclude that it is permissible also when reasoning about integers and variable values as such.

But the true threat of circularity – and the one Poincaré apparently advances – concerns the strength of the principle used and not the subject matter to which one applies it. Bernays' defense is therefore inadequate to meet this threat, for he distinguishes only the ranges of application in the narrow and wide uses of induction and not their deductive strengths.

Hilbert takes on the threat of circularity more boldly than Bernays does on his behalf. In the same [1922] lectures, where he announces that his *Beweistheorie* will uncover a refutation of Poincaré's position, Hilbert stresses that in meta-mathematics complete induction is *not* needed:

We simply have concrete signs as objects, we operate with them, and we make contentual statements about them. And in particular, regarding [an example given of a meta-mathematical proof], I should like to stress that this proof is merely a procedure that rests on the construction and deconstruction of number signs and that it is essentially different from the principle that plays such a prominent role in higher arithmetic, namely, the principle of complete induction or of inference from n to $n + 1$. This principle is rather, as we shall see, a formal principle that carries us farther and that belongs to a higher level; it needs proof, and the proof can be given. (p. 203)

Unlike Bernays, Hilbert at this time does not admit a use of complete induction at the meta-mathematical (what he calls here the lower) level[6], but claims that this principle is essentially part of ordinary mathematics and as such appears in his investigations always purely formally, as an object of study, and never as an instrument of meta-mathematical reasoning. Also unlike Bernays, Hilbert hints at the strength of the principle when he claims that it "carries us farther" than can any contentual reasoning needed for the formal study of proofs. Hilbert thus claims to have refuted Poincaré by describing[7] how the consistency of the principle of complete induction can be secured strictly through the use of principles that in themselves are weaker than it in that they do not carry one as far. Indeed Hilbert's characterization in the passage just quoted of ordinary mathematics as being at a "higher" level than meta-mathematics, whereas usually the opposite image is used, is most likely due to his stressing that the inferential principles of meta-mathematics are weaker, not just in terms of their restricted applicability but in a mathematical sense, than those of mathematics proper.

However, Hilbert's analysis of the situation takes on a different tone towards the end of the 1920s. As his program begins to mature technically he does admit a use of induction at the meta-mathematical level:

[A]s my theory shows, two distinct methods that proceed recursively come into play when the foundations of arithmetic are established, namely, on the one hand, the intuitive construction of the integer as numeral ..., that is, *contentual* induction, and, on the other hand, formal induction proper, which is based on the induction axiom and through which alone the mathematical variable can begin to play its role in the formal system. ([1928], pp. 472–3)

---

[6] For a discussion of the way Hilbert stacks these levels, see note 5 on page 186.

[7] Hilbert announces in these lectures that he has a proof, though he does not present one and, as is known today, in fact could not have.

In contrast to his stance in [1922], Hilbert does not here distinguish what he calls contentual and formal induction other than by saying that only through "formal induction proper" can the mathematical variable receive a recursive treatment. Yet this distinction seems to have more to do again with the proper subject matter of the two principles than with their comparative inferential strength.

With respect to Hilbert's foundational aims, the most that can be made of *this* distinction is that since meta-mathematical induction is part of one's proof-theoretical armory, and since the *Beweisthe-orie* is by design free of all semantic assumptions, this amount of induction should not violate the crucial constraint that one's evaluation of mathematics remain wholly mathematical. Indeed this is Bernays' point in his insistence that meta-mathematical induction never ventures beyond the realm of the concrete-intuitive. As a refutation of Poincaré's challenge, though, this distinction is inadequate. For if one is content to point out that meta-mathematical induction is purely mathematical because of the special ontological makeup of the objects to which it is applicable, leaving open the possibility that it is nevertheless as "complete" in terms of deductive strength as mathematical induction proper, then one has not made any progress towards showing that the proof of the consistency of the principle of induction can proceed in a non-circular way. Rather, one establishes only the disjunctive claim that *either* this proof is non-circular *or* mathematical induction is innocuous enough without proof since it is no stronger than a purely mathematical, semantics-free principle at the heart of proof-theoretical investigations.

This second possibility is just what Poincaré maintained all along, though, and what Hilbert thought was an obstacle to the pursuit of mathematical foundations: that mathematical induction needs no defense because of its especially basic status. According to Poincaré the status of induction is basic in a cognitive sense: It is unprovable because it is already at the heart of thinking and therefore also of proving. (When one keeps in mind Hilbert's ultimate aim of eliminating such ideas from foundational research, it is shocking that Bernays apparently offers the same defense of contentual induction when he says that it is unobjectionable as a tool for analysis because it belongs to "the primitive intuitive mode of cognition.") Speaking on his own behalf, Hilbert describes the sense in which one cannot pursue foundations without induction as methodological: The *Beweistheorie* depends on it. If the induction needed in meta-mathematics is no less "complete" than that used in mathematics itself, though, this difference matters little. Poincaré would find no reason in the results of Hilbert's program thus structured to be dissuaded from his philosophical account of the source of arithmetical truth. Emphasizing the "ontological distinction" between mathematical and meta-mathematical induction thus fails to refute Poincaré's challenge by leaving open this possibility.

But aside from the significance for Hilbert's program of whether meta-mathematical induction is in fact weaker than mathematical induction proper, there is an issue even more crucial for the program's prospects. That is the fact that it was not evident to Hilbert how even to determine whether the two principles differed in strength. As the above passages indicate, he changed his mind about this matter during the course of the 1920s. But if the proof theory is properly mathematical in the sense demanded by Hilbert's aims, should an unambiguous approach to this question not be evident? That one is not is evidence of a lack of rigor undermining the program's principal aims.

The failure of Hilbert's program as it stood in the late 1920s is thus that by isolating a significantly strong form of induction for use at the meta-mathematical level, the proof theorist exempts an awful lot of machinery from non-circular analysis. Bernays, and eventually also Hilbert, thought that avoiding an obscure ontological circularity alone was significant since this was to be avoided precisely by virtue of meta-mathematical induction being restricted to the concrete-intuitive and consequently being purely mathematical. At first sight this might seem like a reasonable enough result for Hilbert in so far as his program amounted to an attempt at a purely mathematical evaluation of mathematics, regardless of the actual strength of the induction needed. However, if the induction is as strong as Hilbert ultimately concedes it is, then at second glance it appears he has attained purity in his methods at the cost of not getting the intended evaluation under way. But the situation is worse even than this: The inability to answer clearly what seems to be Poincaré's main concern – the purely logical circularity in any justification of induction – discloses the ambiguity in Hilbert's work. That his *Beweistheorie* is in fact not purely mathematical is made evident by its inability to distinguish mathematically the two types of induction. What was meant to be a purely mathematical evaluation of mathematics thus turns up both impure and useless for the very evaluations for which it was designed.

So it is not an option for Hilbert to concede that Poincaré was right all along about the impossibility of a non-circular "proof of induction," and rest content that the logically circular proof was at least properly mathematical. For on the one hand this would not suffice to dispel *a priori* explanations like Poincaré's for the grounding of induction, and ultimately all of mathematics. And even more crucially, the inability to determine on the other hand whether such a concession even is necessary reveals that Hilbert had not met his own standard of mathematical purity. Thus in failing to analyze one of his program's chief targets without appeal to the very principle he sought to analyze, Hilbert also failed to cast his analysis in terms explicit enough to allow him to see this clearly.

One must ask finally why between 1922 and 1928 Hilbert felt he had to change his mind about the strength of induction needed at the meta-mathematical level. The most obvious suggestion is that the proofs he sketched in the early part of the decade all turned up erroneous, and further reflection convinced him that stronger methods were needed for conclusive results. But to whatever degree this explains psychologically Hilbert's change of mind, there is a deeper, underlying reason. The second question in Bernays' characterization of Hilbert's program concerned how properly to formulate the statement of consistency. The formulation he settled on was this: "$\perp$ cannot be obtained as the end-formula of a proof." This statement of consistency in terms of the formal, axiomatic development of proofs, though an advance in the development of metatheory, is still informal. For even if one follows Hilbert in treating formulas and proofs mechanically, without attributing any meaning to them or the symbols from which they are built up, a statement *about* proofs and formulas – Bernays' formulation of consistency for example – is itself not such a formula but rather an ordinary statement. To confirm or refute a statement like this, either one's proof theory must be sufficiently informal to treat such statements directly, or to a purely mathematical *Beweistheorie* one must append some extra-mathematical reasoning to determine when informal metatheoretical questions have been settled by mathematical results. The lack of purity in his program thus stemmed from the informality of the consistency statement he aimed to verify. This was the underlying reason why he was unable, in the case of proof theory, to meet the standard he set for himself in his earlier work on the foundations of geometry, namely "to make the discussion of each question of such a character as to ... answer this question by following out a previously determined method and by employing certain limited means." Instead, Hilbert was forced to reconsider his method and means of investigation throughout his program's early history as the insufficiently formal goal proved continuously elusive.

Hilbert thus presented what may properly be thought of as a proto-proof theory, and he took it as far as he could. From within his own circle, the proof-theoretical apparatus never met his own standard of purity and therefore fell short of his goal of demonstrating mathematical autonomy. However, since the research was not adequately pure, its failure does not impugn the original conception of the program. Indeed, by chasing the failure back to the problem of formulation, one learns from this first attempt where to focus for the program's revitalization. For a purely mathematical proof theory, one must first formulate metatheoretical questions in such a way that no reasoning outside of the proof theory is required in order to determine their answers. Only thus can one determine exactly which principles are required in such determinations and thereby learn what can and cannot be proved by non-circular means. And only thus will successful analysis show mathematics speaking for itself.

3.5 HERBRAND'S RECEPTION OF HILBERT

A young student at the École Normale Supérieure named Jacques Herbrand took interest in Hilbert's program in the late 1920s and devoted his doctoral studies to a redevelopment of proof theory according to his own exacting demands of rigor.

Herbrand's colleagues relate in articles and lectures (Chevalley and Lautman [1931] and Chevalley [1934]) written and delivered shortly after his death in 1931 that his ideas were at first unpopular at the École despite his highest reputation among prominent French mathematicians and that he even struggled to organize a committee to supervise his studies. The obstacle he faced was the common opinion among mathematicians that Hilbert's foundational research was essentially a diversion into esoteric philosophical discussion and thus warranted no careful attention from the mathematics community.

In part because of this struggle, Herbrand published some notes while at the École on the nature of Hilbert's program. In these notes he has two main emphases. He points out first that meta-mathematics and proof theory have important and definite mathematical goals that can be studied independently of any philosophical considerations Hilbert associated with them. Second, he suggests that Hilbert's own techniques break down *en route* to establishing the technical goals of meta-mathematics because of a lack of rigor. He proposes, then, to reorganize proof-theoretical technique around a rigorous approach to the purely mathematical problems at the center of Hilbert's program. Herbrand intended these points together as an advertisement for his research: By indicating a theme in Hilbert's program that was isolable from Hilbert's philosophical views and explaining that Hilbert's own methods lacked the precision necessary for the mathematical development of this theme, Herbrand implicitly was making the case for the appropriateness of a dissertation in proof theory in a mathematics department.

Herbrand's efforts paid off. He submitted his thesis in 1930 and in the same year earned a Rockefeller Scholarship allowing him to study in Germany the following year with John von Neumann. During this time he was able to compare his own proof-theoretical contributions with von Neumann's and also to evaluate the impact Gödel's [1931] theorems made on Hilbert's program. That year, inspired principally by his contributions to the foundations of mathematics, Richard Courant reported to Herbrand's father that "[d]uring his short stay in Göttingen and even before this from his works, we have come to respect your son as one of the most promising and, because of the results he had already obtained, prominent young mathematicians of the world" (Chevalley [1934], p. 25). Twenty years later Bernays [1954] would testify to the continuing legacy of Herbrand's thesis, calling his Fundamental Theorem "the central theorem of predicate logic." Clearly his research in proof theory was eventually received as a significant mathematical achievement.

Herbrand's depiction of Hilbert's program is an illuminating backdrop for appreciating fully his own contributions to proof theory. In one note he circulated to accompany his completed thesis, he describes Hilbert's reaction to critics of ordinary mathematics as follows:

The most intransigent critic of mathematical methods could not object to anything in [the techniques to which Hilbert restricts himself] unless he professed that the consideration of a determinate finite number of objects is itself illegitimate; but no one has yet gone that far ...

Hilbert ... raised the problem of resolving the questions [of metamathematics] by the use of exclusively intuitionistic arguments. But, as a result of this, from the start he was led to pose problems such as the following: Consider the axioms of arithmetic. Starting from these axioms and using Russell's rules of inference, we can formulate arguments which Brouwer rejects; however, if we could prove with complete rigor, using intuitionistic procedures, that there is no risk of arriving at a contradiction ..., then Brouwer's critique would lose its point. Thus, we are led to the study of the consistency of the axioms of arithmetic, of analysis, and then of set theory ... These are now well-determined mathematical problems. ...

Hence, we are now faced with the problem of studying, with the method described above, the consistency of all axiom systems imaginable, using only intuitionistic modes of reasoning [i.e., those that Brouwer admits and whose use turns the consistency problem into a determinately mathematical one]. ([1931b], p. 274)

Thus Herbrand makes Hilbert out to be embracing the constraints of Intuitionism and simultaneously undermining Brouwer's criticism by showing that intuitionistic methods suffice to secure the consistency, not only of intuitionistic mathematics, but of all ordinary mathematics. The fact that these constraints turn metatheoretical questions into determinate mathematical problems is a point that Herbrand presents as a mere side effect of Hilbert's methods. This is the exact reversal of the proper understanding of Hilbert's program, according to which Hilbert's primary interest was in presenting metatheoretical questions in a way suited for a fully mathematical treatment free of all philosophical principles. Contrary to Herbrand's analysis, the constraints of finitism (or Intuitionism) were not something Hilbert embraced – they simply were forced by what it takes to reason about signs as such, and it was but a happy accident that the resulting methods allegedly met the demands of mathematics' critics.

Ironically, though, this distorted image of Hilbert's program seems to have helped Herbrand clarify the mathematical nature of his own project. For by sharing the mathematics community's basic misunderstanding of Hilbert, he is able to contrast his own interests against the philosophical ideas associated with the Hilbert school. Herbrand proceeds to explain that his work concerns finding solutions to these "well-determined mathematical problems," and not contributing to or evaluating what he takes to be Hilbert's philosophical theses:

Before indicating the results obtained, we shall mention that Hilbert creates a philosophy of mathematics on the basis of this. One of its fundamental theses, for example, is that once the consistency of an axiom system has been proved ..., then its use is "legitimate" ... But it is not the aim of this work to discuss these ideas and their insufficiency, if any. It attempts only to resolve the mathematical problem mentioned above in the most general case possible. ([1931b], p. 274)

He had made a similar remark already in an essay on the principles of Hilbert's logic that he published in *Revue de métaphysique et de morale*:

We do not claim to explain Hilbert's own ideas in these pages; the instrument he forged is independent of this presentation. We wish only to present the principles of his theory in a form which we shall try to make clearer and less subject to objections than some of those which have been selected up until now. ([1930b], p. 203)

Unwittingly, then, by distancing himself from what he perceived as Hilbert's philosophical ideas and pursuing instead the precisification of proof-theoretical methods, Herbrand would contribute directly to Hilbert's chief philosophical goal, the demonstration of mathematical autonomy. For the problem that he extracted from Hilbert's wide range of foundational writings as an appropriate and important target for mathematical research was the formulation of the consistency of various branches of mathematics as "well-determined mathematical problems" and their investigation "with complete rigor." Herbrand's final summation of his "antiphilosophy" indeed sounds very much like the vision Hilbert had outlined for his program eight years earlier:

We would like to insist again, in conclusion, on the fact that [this new branch of mathematics] is independent of any philosophical opinion; the results obtained are positive; and no more than the mathematician who studies Einstein's equations necessarily partakes of Einstein's ideas must the mathematician who studies the present theories adhere to Hilbert's philosophical principles. ([1931b], p. 276)

However properly mathematical Herbrand considered the fundamental problems of Hilbert's new logic to be, though, he did not consider Hilbert's methods adequately rigorous. He wrote:

At the time these investigations were begun, the status of these questions was as follows: Hilbert had limited himself to giving schemata for proofs, nearly all of which were subsequently seen to be false. Only his proof of

the consistency of the simplest axioms of arithmetic could be sketched in a
somewhat complete manner by his pupil Bernays. ([1931b], p. 275)

The first sign of imprecision thus was the fallaciousness of prac-
tically all of Hilbert's initial attempts at arithmetical consistency
proofs. To Herbrand, though, this fallaciousness was symptomatic
of a deeper error in the framework Hilbert used. In a letter he wrote
to Jacques Hadamard the same year, he elaborated on his misgivings
about the purity of Hilbert's methods as follows:

When I first attacked these questions, the status of them was as follows:
the principal ideas of proofs, and some proof-schemata, had been explained
by Hilbert in a series of articles; but the schemata were almost all false, as
Hilbert himself subsequently recognized. ...It was necessary to consider
the whole theory again from top to bottom in order to attain the desired
rigor in its beginnings and to arrive at precise results ...

    I was obliged to start by establishing all the lemmas, in general easily
proved, which lie at the base of these theories and which had not previously
been stated in a satisfactory manner. I was very often led to complete them.
([1931c], p. 278)

Herbrand thus identifies the source of Hilbert's technical shortcoming as the unsatisfactory statement of the most basic results
of proof theory. In particular he does not think Hilbert had successfully formulated the statement that arithmetic is consistent
as a well-determined mathematical problem, for as we will see
in the next section, his effort at re-patching Hilbert's program
amounts to providing just such a formulation. His appraisal of
Hilbert's achievement accords, then, with ours above. To pick up
where Hilbert left off, Herbrand proposes that meta-mathematical
statements themselves be arithmetized. The primary accomplishment in his doctoral studies is his explanation of how this is
possible.

3.6 INVESTIGATING METATHEORY WITH ARITHMETIC

Recall[8] that Bernays' formulation of the consistency of a theory of arithmetic is the ordinary language statement that no sequence of formulas in the proof calculus for a formalized theory of arithmetic is a valid proof sequence whose last line is the formula $\perp$. The mathematical impurity of this formulation is twofold. First, it is an ordinary statement and not itself a sentence in the language of arithmetic. Second, it is a statement about the existence of certain syntactic objects and the impossibility of certain sequence constructions rather than a statement about the proper subject matter of arithmetic – functions, numbers, and such.

The type of arithmetization most familiar to logicians today addresses the first of these concerns. This is "the arithmetization of syntax," the possibility of which was the monumental discovery of Gödel discussed in Section 3.2. But before Gödel demonstrated how to arithmetize syntax, Herbrand already had addressed the second concern with a different approach to arithmetization. Rather than attempt to find a sentence in the language of arithmetic with the same truth conditions as Bernays' ordinary language statement, as Gödel did, he proposed to substitute for Bernays' statement about syntactic objects a direct expression of consistency in terms of number-theoretical functions. In other words, rather than resolve the impurity in an ordinary language statement of consistency through the arithmetization of syntax, Herbrand formulated an ordinary language statement of consistency that already was "arithmetical." Such a statement is not, like the sentence Gödel would construct, a formula in the language of arithmetic. Its "arithmetical"

---

[8] This section follows the development in Buss [1995] from the definition of strong ∨-expansions to the definition of Herbrand forms.

nature amounts to its verification or refutation consisting in checking purely arithmetical facts (specifically, in solving Diophantine equations) rather than syntactic ones.

Herbrand's style of arithmetization is comparably unknown. It is a crucial contribution to Hilbert's program though, for, as the next chapter explains, Gödel-style arithmetization of syntax on its own cannot salvage Hilbert's program. The following analysis is devoted accordingly to Herbrand's unique approach to formulating metatheoretical claims in arithmetical terms.

According to Herbrand the fundamental problem of metamathematics, in terms of which all other meta-mathematical problems can be worded, is the *Entscheidungsproblem*. Really a class of problems, the *Entscheidungsproblem* is this: given a formal theory of mathematics and a sentence in that theory's language, to determine whether the sentence is a theorem of that theory. Thus, for example, the question of a theory's consistency is the *Entscheidungsproblem* for that theory and the sentence $\bot$. Herbrand considered his Fundamental Theorem a method for reducing each instance of the *Entscheidungsproblem*, as it arises initially as a question about what proofs do and do not exist, to a problem in number theory.

Our statement of Herbrand's Fundamental Theorem involves two uncommon notions, that of a *réduite* of a formula, and that of an "Herbrand proof."

A *réduite* of a formula is defined as follows. First define recursively the strong $\vee$-expansions of a formula: Let Φ be a first-order formula whose only propositional connectives are $\wedge$, $\vee$, and $\neg$ occurring only directly in front of atomic subformulas of Φ. Every formula is a strong $\vee$-expansion of itself, and if X is a subformula of a strong $\vee$-expansion Ψ of Φ whose main connective is an existential quantifier, then replacing X with $X \vee X$ in Ψ produces another strong $\vee$-expansion of Φ.

DEFINITION 1   *A réduite of Φ is the quantifier-free matrix of a prenex-normal form of a strong $\vee$-expansion of Φ.*

Now let T be a first-order theory, and suppose that Φ is a prenex-normal formula in the language of T:

$$\Phi \Leftrightarrow \forall x_1 \ldots x_{n_1} \exists y_1 \forall x_{n_1+1} \ldots x_{n_2} \exists y_2 \ldots \exists y_r \forall x_{n_r+1}$$
$$\ldots x_{n_{r+1}} \Psi(\mathrm{x}, \mathrm{y}). \tag{3.1}$$

A sequence of terms $t_1, \ldots, t_r$ is said to "witness" a formula Φ over T if all the bound variables (if there are any) occurring in each t_i are from $x_1, \ldots, x_{n_i}$ (i.e., the bound variables occurring in t_i all appear to the left of y_i in (3.1)) and $T \vdash \forall \mathrm{x} \Psi(\mathrm{x}, t_1, \ldots, t_r)$. More generally, $t_1, \ldots, t_r$ witnesses Φ (not necessarily prenex-normal) over T *via* Ψ if $\Psi(\mathrm{x}, \mathrm{y})$ is a *réduite* of Φ and $T \vdash \forall \mathrm{x} \Psi(\mathrm{x}, t_1, \ldots, t_r)$.

DEFINITION 2   *An* Herbrand T-proof *of a formula Φ is a réduite $\Psi(\mathrm{x}, \mathrm{y})$ of Φ and a sequence of terms $t_1, \ldots, t_r$ that witnesses Φ over T via Ψ.*

Herbrand's Fundamental Theorem is that T-provability and Herbrand T-provability are equivalent:

THEOREM 3 (Herbrand)   *A first-order formula Φ is valid if, and only if, there is an Herbrand proof of Φ. If T is a universal theory, then $T \vdash \Phi$ if, and only if, there is an Herbrand T-proof of Φ.*

Buss [1995] presents a lucid, proof-theoretical proof of the Fundamental Theorem in this form.

An immediate consequence of Herbrand's Theorem is the "no-counterexample interpretation" of provability, due to Georg Kreisel [1951]. We state it first in a weak form for T a universal theory and Φ an $\exists \forall$-formula.

COROLLARY 4   *Let T be a universal theory. Suppose $T \vdash \Phi$
where $\Phi \Leftrightarrow \exists x \forall y \Psi (x, y, c)$ with Ψ quantifier free. Then for
some $k > 0$ there are terms (all bound and free variables appearing in which are displayed) $t_1(c)$, $t_2(c, y_1)$, $t_1(c, y_1, y_2), \ldots,$
$t_k(c, y_1, y_2, \ldots, y_{k-1})$ such that*

$$T \vdash \forall y_1 \Psi (t_1(c), y_1, c) \vee \bigvee_{i=2}^{k} \forall y_k \Psi (t_i(c, y_{i-1}, \ldots, y_1), y_i, c).$$

Proof.   This case of the no-counterexample interpretation is
straightforward. Since Φ is an $\exists \forall$-formula, its only strong $\vee$-expansions have the form

$$\bigvee^{k} \exists x \forall y \Psi (x, y, c).$$

The prenex-normal form of such an expansion is

$$\exists x_1 \forall y_1 \exists x_2 \forall y_2 \ldots \exists x_k \forall y_k \bigvee_{i=1}^{k} \Psi (x_i, y_i, c).$$

Thus an Herbrand T-proof of Φ is a sequence of terms
$t_1, t_2, \ldots, t_k$ that witnesses Φ *via* the quantifier-free part of this
prenex-normal formula. Such a witness will have the required
variable occurrence constraints.

$\dashv$

A more general form of the no-counterexample interpretation of
provability will be presented below. To state it we need to review
Herbrand's technique of universal quantifier elimination.

DEFINITION 5   *Let $\Phi \Leftrightarrow \forall x \Psi (x, c)$ be a formula with all free
variables displayed. The Herbrand function $h_{\forall x \Psi}$ for $\forall x \Psi$ is implicitly defined by the axiom:*

$$\forall y \forall x (\neg \Psi(x, y) \rightarrow \neg \Psi(h_{\forall x \Psi}(y), y)).$$

Herbrand functions are the dual of the more familiar Skolem functions. Recall that Skolem functions are like choice functions that instantiate existential quantifiers. To foster a similar intuition for Herbrand functions, consider first the obvious validity $\forall y (\forall x \Psi(x, y) \rightarrow \Psi(h_{\forall x \Psi}(y), y))$. Now observe that by rules of passage and contraposition

$$\forall y \forall x (\neg \Psi(x, y) \rightarrow \neg \Psi(h_{\forall x \Psi}(y), y))$$

logically implies

$$\forall y (\Psi(h_{\forall x \Psi}(y), y) \rightarrow \forall x \Psi(x, y)).$$

It follows that

$$\forall y (\forall x \Psi(x, y) \leftrightarrow \Psi(h_{\forall x \Psi}(y), y)).$$

Thus the Herbrand function $h_{\forall x \Psi}$ provides a counterexample to the truth of $\forall x \Psi$ when, and only when, $\forall x \Psi$ is false.

Herbrand proved the equivalence of first-order formulas with their reducts after universal quantifiers have been eliminated with Herbrand functions. More precisely, given a prenex-normal formula $\Phi(c)$, define recursively the *Herbrand form of* Φ denoted $He[\Phi(c)]$ as follows:

1. If $\Phi(c)$ is quantifier-free, then $He[\Phi(c)]$ is $\Phi(c)$.
2. If $\Phi(c)$ is $\exists x \Psi(x, c)$, then $He[\Phi(c)]$ is $\exists x He[\Psi(x, c)]$.
3. If $\Phi(c)$ is $\forall x \Psi(x, c)$, then $He[\Phi(c)]$ is $\Psi(h_{\forall x \Psi}(c), c)$.

Then Herbrand proved $T \vdash \Phi$ if, and only if, $T \vdash He[\Phi]$.

It is possible to restate the version of the no-counterexample interpretation above without universal quantifiers. This statement is an immediate consequence of the earlier one and the result on Herbrand forms just presented:

COROLLARY 6   *Let T be a universal theory. Suppose $T \vdash \Phi$ where $\Phi \Leftrightarrow \exists x \forall y \Psi(x, y, c)$ with Ψ quantifier free. Then for some $k > 0$ there are terms (all bound and free variables appearing in which are displayed) $t_1(c)$, $t_2(c, y_1)$, $t_1(c, y_1, y_2)$, ..., $t_k(c, y_1, y_2, ..., y_{k-1})$ such that*

$$T \vdash \Psi(t_1(c), f(t_1, c), c) \vee \bigvee_{i=2}^{k} \Psi(t_i(c, f(t_1, c), ...,$$

$$f(t_{i-1}, c)), f(t_i, c), c),$$

where f is the Herbrand function for $\forall y \Psi(x, y, c)$.

Thus it is clear why this is called the no-counterexample interpretation of provability. If Φ were not a theorem of T, then given t_i the function f would always return values refuting t_i's claim to witness Φ *via* Ψ. That is, f would provide a counterexample to Φ over T. The corollary states that in case Φ is a theorem of T, the function cannot provide a counterexample: Substituting into some finite number of disjuncts Ψ, taking into account the action of f thus far, eventually results in a witness of Φ over T *via* Ψ.

This statement can be generalized to formulas of arbitrary quantifier complexity. However, since quantifier complexity introduces great variation in the types of a formula's strong $\vee$-expansions, the no-counterexample statement also increases in complexity. Therefore in our statement rather than display subterms we list conditions on them:

COROLLARY 7   *Let T be a universal theory and Φ be a formula with l existential quantifiers. If $T \vdash \Phi$ and Ψ is the quantifier-free matrix of Φ, then for some $k > 0$ there are $k \cdot l$ terms t_{ij} such that*

$$T \vdash \bigvee_{i=1}^{k} \Psi(t_{i1}, f_1(t_{i1}), t_{i2}, f_2(t_{i1}, f_1(t_{i1}), t_{i2}), \dots, t_{il}),$$

where the term t_{ij} can contain as subterms the values of any of the functions f_m, so long as if $m \geq j$ then for no n is f_m applied to t_{in} to generate this value. (For example t_{12} can contain $f_1(t_{11})$ as well as $f_1(t_{21})$ and $f_3(t_{21}, f_1(t_{21}), t_{22}, f_2(t_{21}, f_1(t_{21}), t_{22}), t_{23})$ but not $f_2(t_{11}, f_1(t_{11}), t_{12})$.)

It follows from the deduction theorem that there is an Herbrand T-proof of Φ, just in case for some finite set $\{T_j\}_{j=1}^{l}$ of the axioms of T, there is an Herbrand proof of $\bigwedge_{j=1}^{l} T_j \rightarrow \Phi$. Let us abbreviate $\bigwedge_{j=1}^{l} T_j \rightarrow \Phi \Leftrightarrow P$, and let Π be the quantifier-free matrix of a prenex-normal form of P. In this way one may apply the no-counterexample interpretation to T-provability no matter the complexity of T's axioms: Φ is a theorem of T if, and only if, for some finite number of T's axioms and some $k > 0$, with Π as above there is a disjunction

$$\bigvee_{i=1}^{k} \Pi(t_{i1}, f_1(t_{i1}), t_{i2}, f_2(t_{i1}, f_1(t_{i1}), t_{i3}), \dots)$$

whose terms satisfy the subterm conditions of Corollary 7 that is a propositional quasi-tautology[9] (i.e., a propositional consequence of instances of identity axioms). Thus Herbrand's Fundamental Theorem is a type of reduction of T-provability to propositional provability.

---

[9] In our original definitions for a *réduite* of a formula and for Herbrand proofs, prenex operations are applied only after the formula is strongly $\vee$-expanded. Throughout his writings, Herbrand presents his Fundamental Theorem in roughly the present form and speaks of a *réduite* as a disjunction of multiple copies of the quantifier-free matrix of a prenex-normal formula.

According to Herbrand this is also a reduction of a meta-mathematical problem to an arithmetical one:

The problem consists in trying to make any arbitrary number of propositions $\Pi(t_{i1}, t_{i2}, \ldots t_{in})$ false, or else finding a system of propositions of that form which cannot be made false, because [the Fundamental Theorem] shows that in the first case $[\bigwedge_{j=1}^{l} T_j \rightarrow \Phi]$ cannot be [valid] and in the second case it is.... We are therefore brought back to a purely arithmetical problem, equivalent to the [*Entscheidungsproblem*, for T, Φ]; [if $\phi_j(x_1, x_2, \ldots, x_{n_i})$ are the atomic subformulas of P, then] this problem is the determination of the logical values that must be given to the $\phi_j(t_{j1}, t_{j2}, \ldots, t_{jn_j})$ to make the $\Pi(t_{i1}, t_{i2}, \ldots t_{in})$ false. ([1931a], p. 244, indices and term constants have been changed for consistency)

In a note Herbrand drafted to summarize the results in his thesis, he explains that by reducing the question of the provability of a formula to such a determination, his result "permit[s] the reduction of the most general case of the *Entscheidungsproblem* to the remarkable form of a problem about number-theoretical functions that is but a generalization of the problem of the effective solution of Diophantine equations." As described above, Herbrand considered all questions of meta-mathematics to be instances of the *Entscheidungsproblem*. Thus he continues, "By means of this, all questions which can be raised in metamathematics are 'arithmetized' [*arithmétisées*]" ([1931b], p. 275–6).

To Herbrand the primary value of this arithmetization is not that it paves a way to solutions to instances of the *Entscheidungsproblem*. Though in some special cases it does – an important instance being his own demonstration of the consistency of a weak arithmetic system – the real worth of the reduction is not that it makes the *Entscheidungsproblem* sometimes easy. He explains that the "fundamental theorem poses the [*Entscheidungsproblem*] in an essentially arithmetical form. But we can even [go further and] show that it is a direct generalization of classical problems of arithmetic" ([1931a], p. 251). To illustrate this he considers the questions of the provability of Fermat's Last Theorem in two different arithmetic theories, one with and one without an axiom schema for induction. He explains that the difference between these two questions is that with the introduction of the induction schema one "now has to deal with new functions in constructing" the terms for witnessing the formula. Then he writes:

But the very fact that the general arithmetical problem to which it reduces every question is itself reduced, when we particularize the data, to the solution of Diophantine equations gives us an estimate of the difficulty of the problem ... (p. 253)

That is, the arithmetization in many cases shows clearly that the problem is quite difficult, that "it must be considered ... as beyond the present means of analysis" (ibid.). But for Herbrand the value of arithmetization is evident when a metatheoretic problem is shown to be too hard to solve just as it is when the reduction leads to a solution: In each case a metatheoretic problem has been cast into the precise terrain of mathematics. The crucial contribution to Hilbert's program in Herbrand's thesis is that by exhibiting just which number-theoretic functions one must consider and precisely which equations one must solve, the Fundamental Theorem shows how a question *about* mathematics can be analyzed with mathematics itself. Thus when he remarks (in [1930b], p. 214) that "all efforts in this direction have succeeded only in making the difficulty of the problem more precise," he is not using the term "success" facetiously. The discovery that a problem is too difficult to solve is a small price to pay for its precisification.

3.7 PROGRESS TOWARDS PURITY

Such is Herbrand's standard of rigor. To him a problem is precise only once it falls completely under the purview of mathematical methods. His colleague Claude Chevalley[10] recalled him once saying, "I should like to construct a system that contains all present day thoughts" ([1934], p. 28). One wonders if this attitude discloses more Herbrand's optimism for mathematical systemization or a conservatism about the amount of our mental baggage he counts as genuine thought. Whichever is the case, the investigation of a problem, if it is to be definitive in its findings, must for Herbrand be purely mathematical.

Thus Herbrand saw in Hilbert's failed program a challenge: How can the question of a system's consistency be investigated as a mathematical problem? When Hilbert's *Beweistheorie* proved to be vague in its methods, he and Bernays resorted to a theory about the "givenness" of concrete signs and the admissibility for metatheoretical investigations of reasoning grounded in "the concrete-intuitive" mode of thought. If consistency proofs for theories of arithmetic could be carried out thus, they suggested, then the program could succeed in something of an adulterated form. This "mature" version of Hilbert's program is the one Herbrand was familiar with and the one that he rejected. For resorting to the admissibility of "the concrete-intuitive" mode of thought compromised the possibility of a purely mathematical system of meta-mathematics. He understood Hilbert as trying to show that restrictions to such forms of reasoning were unnecessary since from them much stronger mathematical principles could be verified as

---

[10] Chevalley entered the École one year after Herbrand. He later became a founding member of the *Association des collaborateurs de Nicolas Bourbaki* and an accomplished algebraist.

permissible. But his own reconstruction of the program involved a more wholesale rejection of restrictions on methods, more in line with Hilbert's original vision:

For Herbrand, such restrictions were without foundation, for *he believed that no reasoning whatsoever concerning something given and concrete would be valuable from a purely mathematical point of view*, nor all the more that it was necessary to limit oneself to such reasoning. (Chevalley [1934], p. 27, emphasis added)

His own investigations in proof theory were therefore aimed at eliminating any need to reason about something given and concrete.

When Herbrand takes up the concern of circularity in meta-mathematics, he therefore sounds more like Hilbert in the early 1920s than like Bernays or the later Hilbert. In one place he addresses the question whether the consistency of arithmetic is not already assumed in meta-mathematics. This is a reasonable point to address, for if his main accomplishment has been to turn meta-mathematical problems over to arithmetical methods, then it would seem that one must trust in the consistency of arithmetic in order to investigate problems like the consistency of arithmetic. He expresses this concern in terms of a contrast between his own results and the work of Löwenheim and Skolem (Herbrand considered his Fundamental Theorem to be a constructive version of the Löwenheim–Skolem theorem):

Löwenheim and Skolem implicitly assume that, once we have a model of $\neg P$, P cannot be proved; but this means that the consistency of mathematics (or at least of arithmetic) is implicitly assumed; our theorem, on the contrary, will allow us *to investigate* the consistency of arithmetic. ([1931d], p. 289, emphasis added)

Herbrand's point is that an investigation of arithmetical consistency is not possible if for that investigation the consistency of arithmetic is implicitly assumed. The constructive nature of his theorem is supposed to eliminate such implicit assumptions by disallowing the inference from there existing a model of $\neg P$ to the unprovability of P.[11] But it is too simple to suggest that an unwillingness to make this inference and an insistence on a separate proof of the unprovability of P quiets the threat of circularity, for if for one's proof techniques one needs principles already as strong as those axiomatized in the theory under investigation, then one has only chased circularity out of one's metatheoretical reasoning into the very structure of one's investigation.

Like Hilbert and Bernays, Herbrand is thus at pains to point out that the use of recursive arguments in meta-mathematics is different than the principle of recursion thereby proved consistent:

> It must be carefully noted that there is a certain difference between the manner in which we employ reasoning by recursion here [in meta-mathematical investigations] and the manner in which it is sometimes employed in mathematics [proper]. Here, it is never anything but an indication, in a single formula, of a procedure which must be employed a certain number of times in each particular case. In mathematics, on the other hand, it can happen that this reasoning is used in the case of concepts for which material representation is not possible, as opposed to our signs; as, for example, for the set of integers or of real numbers. That is why it will not be tautologous for us to prove later on [with the aid of meta-mathematical recursion] that reasoning by recursion, in its use in arithmetic, is consistent. ([1930a], p. 51)

Unlike Bernays and eventually also Hilbert, though, Herbrand does not phrase this distinction simply in terms of the subject matter

---

[11] Herbrand describes his theorem as making this intuitive inference unnecessary as follows: "Consider an axiom system dealing with certain objects, and assume that a model of these axioms can be constructed. That is, a set of objects can be constructed in which, by means of suitable definitions of the different relations and functions which appear in the axioms, the axioms are true. Then the theorem in question amounts to the assertion that the axioms system cannot be inconsistent. But all this must be translated into a statement satisfying the intuitionistic rules of reasoning, and using only the properties of the signs of our symbolic language" ([1931b], p. 275).

attended to in the two uses of induction. Rather, he points out that recursion in meta-mathematics "stops in the finite" – i.e., that it is always applied to a single formula at a time and only in a definite number of steps ([1930a], p. 50). This is the kind of "recursion" that Bernays distinguishes from "complete induction" when he remarks about it that "we cannot get by in this way alone." The recursion that is to be proved sound, by contrast, is applicable to infinite totalities like the set of all integers. Herbrand does characterize this distinction in terms of the impossibility of the material representation of the concept of the complete set of integers. But unlike Bernays he is not trying with this characterization to argue his way out of circularity by distinguishing the type of subject matter to which the two principles (mathematical and meta-mathematical induction) are applicable. The impossibility of a concept's material representation is relevant to Herbrand because it is a sign of its infinitary nature, and it is the possibility of grounding an induction principle applicable to infinite totalities with only finite recursion that he cites as evidence of his investigation being earnest and gainful.

As a consequence Herbrand can claim to have made two advances in proof theory. First, if he is correct, then he has been able to solve a problem that Hilbert had announced as one of his program's targets but then never conclusively solved, namely, the refutation of Poincaré's claim that mathematical induction could never be justified without its use in its own justification. And more importantly he has made meta-mathematics precise enough to demonstrate conclusively exactly where and what strength of induction is used in meta-mathematical proofs. Thus, even if he is wrong on the first point, then at least this error can be conclusively verified.

Nonetheless, Herbrand's results fall short of mathematical purity. The formulation of consistency with which Hilbert and

Bernays worked presented two obstacles to their program. One was that, as a statement about syntax, a mathematical verification of the statement was *prima facie* not possible. Without a mathematical theory of syntax, keeping track of the principles used in one's reasoning about formulas and signs would always involve an extra-mathematical step, an appeal for example to a semantic theory. When this became evident to Hilbert, he bit the bullet, argued for the innocuousness of this much extra-mathematical armory, and in effect gave up on foundational problems like Poincaré's that he could not settle conclusively with his *Beweistheorie*. Herbrand, by contrast, overcame the challenge by showing how to reformulate consistency in arithmetical terms, so that the verification or refutation of his formulation of consistency was merely a matter of solving or proving unsolvable the appropriate number-theoretical equations.

But the second obstacle in the way of Hilbert's program persists despite Herbrand's achievement. To see this consider the statement of consistency one generates with the Fundamental Theorem:

A theory T is consistent if for no finite set $\{T_j\}_{j=1}^{l}$ of its axioms is there a $k > 0$ and terms t_{ij} satisfying the conditions of Corollary 7 such that a disjunction of the form

$$\bigvee_{i=1}^{k} X_T(t_{i1}, f_1(t_{i1}), t_{i2}, f_2(t_{i1}, f_1(t_{i1}), t_{i2}), \ldots),$$

is a quasi-tautology, where X_T is the quantifier-free matrix of $He\left[\bigwedge_{j=1}^{l} T_j \rightarrow \perp \right]$.

Although this statement is verifiable with arithmetical means, the statement is not in the language of arithmetic. Consequently, even after the appropriate mathematical problem has been solved – after for example the Herbrand functions f_j are shown always to provide a counterexample no matter which axioms are chosen and no matter what term substitutions are made — one is not left with a theorem in arithmetic to that effect. What Herbrand did not provide, then, was an entirely arithmetical sentence that is a theorem if, and only if, T is consistent. Herbrand's formulation is an improvement on Bernays' because it has mathematical verification conditions, but to announce *that* the statement has been verified one must still step outside of mathematics.

To overcome this final hurdle, one must be able to express such phrases as "for some conjunction of T's axioms" and "for all sequences of terms" directly with mathematical formulas. No one proposed a way to do this until the year after Herbrand completed his thesis, when Gödel showed how to arithmetize syntax. Gödel's work does not build on Herbrand's so much as bypass it, though. For the formulation of consistency that he "arithmetizes" is the one used by Hilbert and Bernays. With respect to mathematical autonomy, then, Gödel's arithmetization of consistency is wanting for the very opposite reason that Herbrand's is: The sentence representing the consistency of a theory might be verified or refuted with an arithmetical proof of it or its negation,[12] without it being possible to determine by purely arithmetical means that it is a statement of consistency. Thus neither in Herbrand's work nor in Gödel's does one find an adequate solution to Hilbert's problem of formulation.

Finally it is worth pausing over the remarkable fact that Herbrand recognized *both* inadequacies in the formulation of consistency used by Hilbert and Bernays. Herbrand recognized that his "arithmetical" formulation of consistency still is not itself a formula in the language of arithmetic and thus suffers from a lack of

---

[12] Or, as in Gödel's example for the theory of *Principia Mathematica*, shown to be undecidable.

purity. He called attention to the problem in the introduction of his dissertation. There he explained that "[t]he properties to which we shall apply reasoning by recursion [in meta-mathematics] are such that one can always recognize whether or not they hold with regard to any particular proof or proposition (for example, the property of containing, or not containing, a given sign)." However, he found our ability to recognize when a property is satisfied to be inadequate if it cannot be shown mathematically which criteria this recognition rests on:

> It is certain that a criterion enabling us to recognize if a property satisfies this condition cannot be given [with present techniques], though in each particular case that this is so it is easily seen to be true.[13] *To avoid this appeal to intuition, it would be necessary to do work for this logic [i.e., for the informal proof theory] analogous to that which permits the axiomatization of ordinary mathematics*; but this is perhaps yet premature. ([1930a], p. 50)

Apparently, then, Herbrand distinguished the type of arithmetization developed in his dissertation and the type Gödel would introduce, understood that his development is strictly of the first type, and recognized a need for new methods directed at the arithmetization of syntax, without which meta-mathematics remains plagued by occasional "appeals to intuition" disrupting it as a fully rigorous science – all a year prior to Gödel's results. Demonstrating mathematical autonomy was not among Herbrand's explicit goals, nor did he understand it as the target of Hilbert's program. His infatuation with rigor nonetheless led him to see precisely what was needed to repair Hilbert's program and to measure accurately his own partial contribution to that reparation.

---

[13] This sentence has been retranslated as closely as possible to Goldfarb's ungrammatical rendering: "It is certain that a criterion enabling us to recognize if a property satisfies this condition cannot be given, but in each particular case that this is so will easily be seen to be true."

CHAPTER 4

Intensionality

4.1 INTRODUCTION

In [1960] Solomon Feferman observed a philosophical distinction between Gödel's two incompleteness theorems (Gödel [1931], *Satze* 6, 11). The first of Gödel's theorems is that if the system S comprised of the Dedekind–Peano axioms, the underlying logic of Russell's *Principia Mathematica*, and the axiom of choice is ω-consistent, then it is also *incomplete* (i.e., there are sentences Φ in the language of S such that, if S is ω-consistent,[1] then neither Φ nor $\neg\Phi$ is S-provable). Gödel proved this by exhibiting, as an instance of such a Φ, the arithmetization of the metatheoretical statement that declares its own S-unprovability. The second theorem says that another unprovable sentence is the arithmetization of the metatheoretical statement of the consistency of S. Feferman's distinction is not strictly speaking one between these theorems, but rather between the way they are ordinarily paraphrased in natural language. The first theorem is taken to mean that if S is consistent, then there are sentences in the language of S that S can neither prove nor refute. Feferman finds this unproblematic, in contrast to the typical rendering of the second theorem

---

[1] Rosser [1936] strengthened the theorem by reducing the hypothesis of ω-consistency to that of plain consistency. Hereafter "Gödel's first incompleteness theorem" will be the name given to Rosser's strengthenings thereof.

as, "If S is consistent then it cannot prove its own consistency."[2] By understanding the second theorem this way, he claims, one implicitly holds some complex commitments about the nature of Gödel's arithmetization techniques. Let us close in on Feferman's distinction in three stages.

4.1.1 First pass

The main thrust of Feferman's distinction is simple. Although Gödel proved his first incompleteness theorem by exhibiting a specific arithmetical sentence that he took to express its own unprovability, nothing in the standard reading of the theorem turns on the details of, nature of, or veracity of this "expression." Supposing that the sentence really expresses nothing at all, that it is (as it appears to be when fully displayed) a meaningless mathematical formula, it is still a sentence in the language of arithmetic and is only provable in that arithmetic if it is also refutable. Under this supposition the fact that the sentence is an arithmetization of the odd statement declaring its own unprovability merely served as a helpful aid to Gödel. Since an analogous sentence declaring its own falsehood had been known already to the ancient Greeks to be paradoxical, Gödel was able to apply in the setting of arithmetic the same reasoning used to determine why that sentence could not coherently be thought of as true or as false to show that a formal expression with its same basic structure was neither provable nor unprovable. The fact that this worked is evidence only that arithmetization is a useful way to construct

---

[2] The consistency of S suffices for the arithmetization of the metatheoretical statement of the consistency of S to be unprovable. Throughout this book, I follow standard practice by referring to results about the unprovability of these "consistency formulas" as versions of Gödel's second theorem. To show further that these consistency formulas are, like the Gödel sentence, undecidable, one also needs to know that S is ω-consistent.

sentences with desired proof conditions, not that the mathematical formula means the same thing as the sentence it arithmetizes. By contrast, if the above paraphrase of Gödel's second theorem is accurate, then the sentence Gödel proved to be unprovable if S is consistent must actually express the consistency of S.

One might attempt a glib dismissal of this distinction. The mathematical content of Gödel's theorems is not ambiguous. Therefore since the theorems do not say anything about "expression," one might argue, the distinction Feferman presents is about a human tendency to paraphrase carelessly. What is philosophically interesting about that? This dismissal will not work, though. Consider Gödel's own summation of his second theorem: "Hence a consistency proof for the system S can be carried out only by means of modes of inference that are not formalized in the system S itself [assuming S's consistency], and the analogous results hold for other systems as well" ([1931], p. 596). Two points come to light. First, to dismiss Feferman's distinction is to call Gödel careless. The paraphrase of Gödel's theorem under scrutiny is not the lay-person's muddled reading of the result; it is the theorem's intended content. Feferman's point is that something more is needed to determine whether Gödel demonstrated what he was after in his second theorem than what is needed to make the same determination for his first theorem. Second, there is nothing fancy about the type of "expression" being debated, so it will not do to say simply that Gödel could not have demonstrated what he was after since a mathematical theorem can say nothing about expression. Gödel took himself to have proved that any proof of the consistency of S would have to utilize principles not formalized in S. Feferman is pointing out that something complex is involved in inferring this from Gödel's second theorem.

4.1.2 Second pass

Demystifying the notion of expression behind the paraphrase of Gödel's second theorem leads to a more precise understanding of the role arithmetization plays. One wants to know simply whether a proof of the consistency of S can be formalized in S. Gödel's arithmetization of the statement that S is consistent then qualifies as an adequate expression of S's consistency if one can infer from his second theorem that no proof is formalizable thus. On the one hand, one presumably need not defend any theoretical account of semantic content in order to determine whether this inference is admissible. That is the sense in which the notion of expression has been demystified. But for the very same reason a new task arises. One cannot simply validate Gödel's arithmetization techniques by showing that they accord with some theoretical account of meaning. At most that would show that no proof of the particular formulation of consistency that Gödel arithmetized is formalizable in S. There would remain the possibility that a proof of some other formulation of S's consistency could be embedded in S.

Gödel did not think of his theorem as restricted to particular formulations of provability and consistency. By considering it stable grounds for believing that no proof of the consistency of S could be formalized in S, he meant that the details of the formulas appearing in his proof were unimportant, that the upshot of the theorem was fully general. But then the question of adequate expression reappears, for how is one to be confident that the theorem has such general consequences?

This question has been emphasized by Michael Detlefsen. He argues that the second incompleteness theorem may not be such stable grounds for the conclusion Gödel draws from it:

The Stability Problem is, then, precisely this: to show that every set of properties sufficient to make a formula of T a fit expression of T's consistency is also sufficient to make that formula unprovable in T (if T is consistent). ([1986], p. 81)

Detlefsen uses the term "expression" differently from how we have been using it. For us a T-unprovable formula would not be a fit expression of consistency unless one could infer from its unprovability that no proof of T's consistency can be carried out in T. That is what we mean by saying that a mathematical formula expresses consistency. Detlefsen's terminology allows for there to be multiple formulas in the language of T all adequately "expressing" consistency, some provable in T and others unprovable in T. His Stability Problem still makes sense in our terms, though: How can one show that knowing of the unprovability in T of a mathematical formula suffices for us to know that the consistency of T is unprovable in T by any means?

Thus Feferman's distinction comes into sharper focus. The point of the first incompleteness theorem just is the incompleteness of arithmetic. Nothing more must be known about the formula Gödel constructed for this theorem than that it is neither provable nor refutable in S. Even if one wants to understand the theorem in the robust form that there is an arithmetical truth unprovable in S, thereby adding to the theorem's mathematical content a claim about the truth of the Gödel sentence verifiable only outside of S, one still need not be concerned with whether that sentence actually expresses its intended content. But to justifiably believe that S's consistency cannot be proved in S, one must know that Gödel's arithmetization of consistency is of such a nature that its S-unprovability testifies to the impossibility of *any* proof of S's consistency being embedded in S.

4.1.3 Third pass

Suppose that one could formulate a statement of S's consistency, the standard encoding of which generates a mathematical formula provable in S. According to Feferman, Gödel's arithmetization of consistency does accurately express the consistency of S, yet he finds nothing paradoxical in this supposition. In fact he presents an encoding of a formulation of S's consistency and shows that its proof in S is trivial. His explanation for how this does not jeopardize the claim that the formula Gödel constructed accurately expresses S's consistency – that consequently the second theorem rules out the formalization in S of any proof of S's consistency – brings his distinction fully to light.

According to Feferman when one asks whether a theory T can prove its own consistency, one wants to know if there is a sentence recognizable in T as a statement of T's consistency that T can prove. He claims that a proper analysis of Gödel's second theorem (for at least most consistent arithmetical systems T) shows that the answer to this question is "no." Therefore no theorem of T can be understood in T as a statement of T's consistency. In particular if an arithmetization of some formulation of T's consistency is a theorem of T, then the resources of T are inadequate to determine that the theorem is a statement of T's consistency.

Feferman does not argue for this point. He takes it to be obvious. If some proof of T's consistency can be formalized as a T-proof of an arithmetization of T's consistency, but T does not see the proved formula as a statement of its own consistency, then it would be incorrect to describe this situation as T having proved its own consistency. One could say that there is a proof in T of a formula that arithmetizes T's consistency, but not that there is a proof in T of T's consistency. (Oedipus knew that on his return from the Oracle he slew the unarmed king of Thebes, but not until the seer Tiresias later disclosed the crucial identity did he know he had confirmed the Oracle by killing his own father.) "[T]he applications of the method [of arithmetization]," writes Feferman, "can be classified as being *extensional* if essentially only numerically correct definitions are needed, or *intensional* if the definitions must more fully *express* the notions involved, so that various of the general properties of these notions can be formally derived" ([1960], p. 35). He claims that incompleteness results require only arithmetization of the extensional type, but a result about the unprovability of consistency must be intensional. That is, in order for results about the unprovability of a formula to have the intended metatheoretical content, the formula's expression of the consistency of T must be explicit, in the sense that T recognizes it as such a statement.

Notice that this indicates a solution to the Stability Problem discussed above. Whether or not an arithmetization of the statement that T is consistent is intensionally correct is determined strictly by what properties T proves about the arithmetical formula. If T fails to prove that all the salient properties hold for some formula, even if it is a numerically correct encoding of a formulation of T's consistency, then a proof in T of this formula will not be a T-proof of T's consistency. Thus if all formulas for which T proves those salient properties are unprovable in T, then so too is T's consistency. The Stability Problem gets its sting from the presupposition that multiple formulas all expressing the consistency of T could be inequivalent in T. If one thinks of the expression of consistency intensionally, though, then the sting goes away. From the intensional point of view, no formula can express T's consistency without T knowing about it, and it is odd to think of T seeing its own consistency equally expressed by formulas it cannot prove to be equivalent.

The difference between the two incompleteness theorems Feferman points out can thus be put concisely: To show that an arithmetical system is incomplete, only numerically correct arithmetization is needed, but to show that the system, if it is consistent, cannot prove its own consistency, the arithmetization used must be intensionally correct.

4.2 GÖDEL'S SECOND THEOREM AND HILBERT'S PROGRAM

In what follows I analyze Feferman's argument for why Gödel's arithmetical consistency statement is intensionally correct and develop a notion of intensionality applicable to the treatment of metatheory for mathematical theories generally. But before taking up these tasks I must clarify how intensionality bears on the prospects for a revival of Hilbert's program.

Gödel's second theorem, as explained above, is usually understood to be a proof that no sufficiently strong, consistent mathematical theory can prove its own consistency. As discussed in the previous chapter, this understanding is commonly taken to be a refutation of Hilbert's program. The reasoning is that Hilbert wanted to prove the consistency of ordinary mathematics in a non-circular way, and finding that the techniques required for a consistency proof must extend the techniques formalized in the system being proved consistent aggravates this ambition by showing that circularity is unavoidable.

Two ways out of this failure have been suggested. Ironically the first one is due to Gödel. In the paper where he announced his incompleteness results, he writes:

I wish to note expressly that [the second incompleteness theorem] ...do[es] not contradict Hilbert's formalistic viewpoint. For this viewpoint presupposes only the existence of a consistency proof in which nothing but finitary means of proof are used, and it is conceivable that there exist finitary proofs [of the consistency of S] that cannot be expressed in the formalism of P [$=$ *Principia Mathematica*]. ([1931], p. 615)

Recall that for Hilbert, finitism is whatever it takes to reason effectively about formal proofs as such. Thus it was ambitious to predict that what would be needed for the *Beweisetheorie* would be formalizable in S, for this would allow a gainful mathematical evaluation of arithmetic, but there was no guarantee that formalization in S would be possible. According to Gödel this prediction was *too* ambitious, for he took himself to have shown that what one needs to demonstrate the consistency of S extends the principles formalized in S. Gödel only hesitates to draw the further conclusion that there is no finitary consistency proof of S. Perhaps it takes more to reason effectively about signs than Hilbert guessed it would take, he thought, but that amount of reasoning still could be finitary.

Gödel's proposal ends in the same dilemma that Hilbert faced when considering Poincaré's challenge, though. For Gödel's theorem obviously holds for all mathematical systems stronger than S as well. Moreover, if the second theorem for S rules out the formalizability in S of any proof of S's consistency, then the analogous consequence holds for all theories T extending S. At the 1931 meeting of the Vienna Circle – the first such meeting following Gödel's results, Gödel quoted John von Neumann reflecting on this situation as follows: "If there is a finitist consistency proof at all, then it can be formalized. Therefore, Gödel's proof implies the impossibility of any [finitist] consistency proof [of all of mathematics]." To this Gödel replied that the formalizability of all finitary reasoning is not so obvious: "[T]he weak spot in Neumann's argumentation," he claimed, is that it is unclear whether all finitary proofs could "be captured in a *single* formal system" (cf. Sieg [1986], p. 342). If they cannot be, then the consistency of all ordinary mathematics could proceed in a series of proofs, no one of which is embeddable in the fragment of mathematics to which it applies, but each of which is finitary. In this way only finitary means would be needed for the verification of the consistency of all mathematics. Gödel's "charitable" stance is not helpful, however, because although it admits the possibility of a finitary proof theory, it does not allow for a proof theory that ever could investigate the consistency of a mathematical system without implicitly assuming the validity of the very principles under investigation. This "way out" is unwelcome to the Hilbertian because it leaves no mathematics open to non-circular analysis.

The second proposal for rescuing Hilbert's program from its apparent refutation is Detlefsen's claim that the second incompleteness theorem for a theory T is not stable grounds for the conclusion that there is no proof in T of T's consistency. According to Detlefsen, even the most general versions of Gödel's theorem do not rule out the possibility of there being a T-provable formula expressing T's consistency. As evidence he describes how to formulate a notion of T-provability that is equivalent to the formulation Gödel uses only if T is consistent in Gödel's sense, argues that his formulation is acceptable on certain "instrumentalist" principles that he identifies with Hilbert's theory of knowledge, and demonstrates that the arithmetization of this formulation of provability does not satisfy the derivability condition "formalized local provability completeness":

$$\text{For all formulas } \Phi \text{ of the language of } T,$$
$$T \vdash Thm_\tau(\overline{\Phi}) \rightarrow Thm_\tau(\overline{Thm_\tau(\overline{\Phi})}).$$

Since only statements of consistency in terms of predicates satisfying all the derivability conditions are shown to be underivable by Gödel's second theorem, that theorem leaves open the question whether the consistency statement based on this formulation is provable.

Detlefsen thus describes a way for T to prove an arithmetization of a nonstandard formulation of its own consistency, and he argues that this formulation should be acceptable to Hilbert because of his instrumentalist philosophical views. His proposal is problematic for three reasons, though.

First, the formulation of provability Detlefsen describes does not guarantee that all T's theorems (ordinarily construed) count as "actual" theorems (as determined by the new formulation). Detlefsen explains that "for the instrumentalist ... discovery of an inconsistency" can be reconciled in the following way:

In one proof, axioms $A_1, \ldots, A_n$ and rules $R_1, \ldots, R_n$ might lead to the efficient generation of a real truth, whereas in another they might lead to a contradiction. So long as the instrumentalist has a way of isolating the former from the latter, there is no need for him to declare both proofs illegitimate. ([1986], p. 121)

The instrumentalist may instead ascribe genuine proof-hood only to the former and eliminate the latter "would-be proof" from his lexicon. Consequently the mathematics for which one might generate a consistency proof by following Detlefsen's procedure may be only a fragment of ordinary mathematics. Since Hilbert intended to validate all ordinary mathematics and demonstrate that restrictions on mathematical practice were unwarranted by foundationalist considerations, this consequence compromises the point of his program.

Second, if Detlefsen's proposal salvages a research program Hilbert would have endorsed, it would have been only as a weak consolation for the program he ultimately was after. Detlefsen's proposal turns entirely on securing a foundation of mathematics in keeping with Hilbert's instrumentalist epistemology:

> [S]ince the Hilbertian only advocates the use of such ideal proofs as would yield a gain in efficiency when used in lieu of their contentual counterparts, it follows that (strictly speaking) it is only such proofs whose real-soundness he must establish. (Detlefsen [1986], p. 86)

It is a matter of debate whether Hilbert's general epistemic outlook was instrumentalist or realist,[3] but that issue is unrelated to the direction Hilbert's program should take. Hilbert's principal goal was to show that mathematics could stand on its own feet by providing its own answers to questions about whether and how its methods work – and that it could do this in a way that would be acceptable to everyone because it favored no epistemological views, including whatever views Hilbert entertained. A program compelling only to those who espouse a certain form of instrumentalism is precisely the kind of foundational program Hilbert set out to show was mistaken and a threat to the privileged status of mathematics among the sciences.

The final reason Detlefsen's proposal is unsatisfactory is that he does not consider whether any arithmetical sentences whose T-provability is not ruled out by Gödel's result are recognizable as expressions of consistency *by* T. Even if we were to grant the legitimacy of formulations of consistency based on instrumentalist principles, that is, what guarantee would we have that their arithmetizations would be recognized by T as stating that T is consistent? Detlefsen does not discuss the issue, but rather rests content with the claim that the variant notion of provability he introduces

---

[3] Kitcher and Mancosu agree with Detlefsen's ascription of an instrumentalist epistemology to Hilbert, while Prawitz, Simpson, and Hallett disagree. See Mancosu [1998a] for a discussion.

is acceptable from an instrumentalist point of view and the argument that Hilbert endorsed the instrumentalist principles required to agree about this.

As discussed in the previous section, Detlefsen's Stability Problem appears most urgent when one views the arithmetization of consistency statements purely extensionally. For it is according to this conception that the possibility of multiple formulas, all expressing T's consistency but not all T-provably equivalent, is plausible. Perhaps Detlefsen ignores the issue of intensionality because the demand for explicit expressions of metatheoretical properties deflates the Stability Problem, which from his point of view is so evident. As a result, his proposal ends up in a position worse than Gödel's.

Just as Gödel's suggestion left open the possibility of a finitary consistency proof of all mathematics at the cost of not allowing for a non-circular analysis of any mathematics, a purely extensional development of arithmetization prevents consistency proofs ever from embedding fully into the systems they are about. One has to step out of the system in order to complete such proofs – in order to see, that is, that they are consistency proofs. It is essential to Hilbert's program that the recognition of proofs as consistency proofs be done within proof theory itself, to insure that one need not appeal to extra-mathematical grounds in determining what the proved formulas express. Thus since an extensional development of arithmetization leaves over this determination to be made outside of the system, the proof theory must extend the system in order to incorporate the principles needed to verify the arithmetization. In this way Detlefsen's instrumentalist consistency proofs would be circular, just as would be Gödel's finitary ones.

The prospects for Hilbert's program are actually even bleaker in the game of searching out solely extensional counterexamples to the stability of Gödel's result than in conceding that the theorem rules out the formalizability of any proof of T's consistency in T. The ability of T to recognize one of its own formulas as a consistency statement is not just one among many steps in T proving its own consistency. For Hilbert's purposes it is the most important step. In fact, even if Hilbert's program cannot be carried out in its ideal form, with non-circular proofs of consistency everywhere they are called for, the ability to put the question of a theory's consistency to that same theory is by itself of great significance to Hilbert. For the primary challenge Hilbert sought to overcome (and the one that Herbrand took up after him) was that of formulating metatheoretical questions purely mathematically and still meaningfully, i.e., as mathematical problems in settings where their solutions are not implicitly assumed. Although Gödel's analysis of his own theorem convinced him that non-circular consistency proofs are not possible, that is only because he saw his arithmetical consistency formula as a full enough expression of S's consistency for its unprovability to entail this. Thus, unlike Detlefsen, he acknowledged that Hilbert's problem of formulation can be solved and even believed that he had solved it.

Although neither of the proposed fixes for Hilbert's program describes a way to revive all of what Hilbert wanted from his foundational endeavor, the proposals themselves, in the problems they pose and the possibilities they entertain, make clear exactly what role intensionality must play in a development of metatheory according to Hilbert's principle of methodological purity. Therefore one need not agree with Feferman that the notion of the T-provability of metatheoretical properties is, like the epistemic notions of knowing and believing, inherently intensional to appreciate the need for an explicit arithmetization of metatheory. Mathematical autonomy demands it.

4.3 FEFERMAN'S APPROACH

Since Gödel-style arithmetization involves representing essentially non-mathematical claims (claims about syntax) with arithmetical formulas, the question immediately arises how an arithmetical system on its own can verify the accuracy of that representation. The system has the task of verifying that one of its formulas properly captures a notion that it cannot directly express. One seemingly can make this verification only from within a background theory where both the arithmetical claim and the syntactic claim are expressible, which would be a setting where arithmetization is not needed: One should simply prefer to work directly with the syntactic formulation. But of course this is just what the arithmetical system cannot do, so the prospects for an "internal" verification look dim.

Feferman's approach to this question is to give up on any attempt at a verification of the representation, "internal" or "external." An internal verification is *prima facie* impossible, and an external verification is useless for establishing claims about what a mathematical system can and cannot prove about itself. Rather, in order to demonstrate that a formula expresses a metatheoretical notion, one checks that the system proves that the formula has certain properties that are essential to or constitutive of that notion.

For example, Feferman considers primarily the arithmetization of axiom-hood. He finds that there is no uniform answer to the questions which numerical representations of the predicate "x is an axiom of T" are intensionally correct, and he submits this finding as a cautionary note about reading too much into standard results for theories like *Peano Arithmetic*. Specifically he found (corollary 5.10) a binumeration α^* of PA that is extensionally correct, but whose corresponding consistency statement is provable in PA. One might hastily conclude that the arithmetization that gives rise to such an enumeration of PA blocks any generalization of Gödel's second theorem from the specific arithmetization of Gödel [1931]. But Feferman draws the opposite conclusion: While the generalization of the technical result is restricted[4] by this variant binumeration, the generalization of the "unprovability of consistency" is not:

> We have maintained that insofar as a formula α expresses membership in A, the formula Pr_α expresses provability of $\mathcal{A}$ in $\mathcal{M}(\mathcal{P})$ and the sentence Con_α expresses the consistency of $\mathcal{A}$ in (M) and (P). Thus, one particular conclusion we can draw is that the formula α^*, although it extensionally corresponds to A, does not properly express membership in $\mathcal{A}$. ([1960], p. 69)

The expression of membership in a theory must be intensional in order for it to be a proper expression on top of which one may formally define metatheoretical properties. That is, the failure of Gödel's second theorem in some general settings is independent of the stability of the result on the unprovability of consistency when the settings in question are ones where intensional arithmetization fails.

The question of intensionality, on Feferman's analysis, thus reduces to the question of which properties are constitutive of the various metatheoretical notions so that the provability in T that a formula has these properties amounts to a demonstration of that formula's genuine expression of the property being arithmetized. Feferman thinks that this question can be further reduced to asking for which encodings of T's axioms the usual constructions of formulas for provability, consistency, etc., satisfy these properties. For

---

[4] Feferman's study revealed that the result must be restricted to recursively enumerable representations of theories, though it holds universally for those arithmetizations (Theorem 5.6 of Feferman [1960]).

example Feferman requires an intensionally correct arithmetization of T-provability to satisfy

$$T \vdash \forall u \forall v \forall w (Fmla(u) \wedge Term(v) \rightarrow Fmla(\overline{sub(u, v)}))$$
$$T \vdash \forall \overline{\phi} \forall \overline{\psi} (Thm_\tau(\overline{\phi}) \wedge Thm_\tau(\overline{\phi \rightarrow \psi}) \rightarrow Thm_\tau(\overline{\psi}))$$
$$T \vdash \forall u (Proof_\tau(u) \rightarrow Thm_\tau(\overline{Proof_\tau(u)}))$$
$$T \vdash \forall \overline{\phi} \forall u (Prf_\tau(u, \overline{\phi}) \rightarrow Thm_\tau(\overline{Prf(u, \overline{\phi})}))$$
$$T \vdash \forall \overline{\phi} (Thm_\tau(\overline{\phi}) \rightarrow Thm_\tau(\overline{Thm_\tau(\overline{\phi})}))$$

One can thus say concretely why the variant definition of PA discussed above fails to admit an intensionally correct arithmetization of PA-provability. Although it is numerically correct, the provability predicate one builds from it fails some of the above conditions, so that PA does not see in this predicate the notion of provability.

An obvious question to ask about Feferman's notion of intensionality, upon seeing that not all extensionally correct representations of being an axiom of T are intensionally correct, is whether the converse holds. Feferman discusses this briefly and points out that some intensionally correct arithmetization schemes could fail to accurately binumerate the intended properties, so that the answer is "no." Thus the intensionally correct arithmetizations are not just a subclass of the numerically correct ones. This is an important observation. Feferman is not trying to refine the notion of arithmetization in order to pick out just those binumerations that are genuine expressions of the intended notions. He proposes that one ignore numerical representation altogether by pursuing arithmetization explicitly within T without considering whether T's treatment of its own metatheory corresponds with the way that T's metatheory appears from a richer standpoint.

One might take Feferman's proposal as a defense of the Hilbert–Bernays–Löb derivability conditions for the second incompleteness theorem. If only intensionally correct arithmetizations are admitted, then there is no question about the meaning of Gödel's second theorem for PA. Any candidate arithmetization that skirts the derivability conditions does so by failing to produce an intensionally correct arithmetization of theorem-hood. Therefore even if the resulting consistency statement were a theorem of PA, this could not count as a proof in PA of the theory's consistency. Intensionality criteria are a set of necessary conditions for an arithmetization to meaningfully represent metatheory in a purely arithmetic environment. Derivability conditions are a set of conditions on a provability predicate sufficient for Gödel's second theorem. Since the Hilbert–Bernays–Löb derivability conditions are consequences of Feferman's intensionality criteria, one may infer from Gödel's second theorem the unprovability of a theory's consistency. But as Feferman points out, there is no need to impose the derivability conditions as additional criteria and verify that the specific encoding with which one works satisfies these. One simply can say that any arithmetization of consistency, if it is intensionally correct, will be unprovable without exhibiting any particular encodings.

Feferman does not argue for the accuracy of the intensionality criteria he chooses. In fact he does not even list criteria that he claims are necessary or sufficient conditions on an arithmetization for it to be intensionally correct. One could reasonably demand a defense of the above criteria though, on the grounds that they seem from one point of view rather strong and from another point of view too weak. It has been suggested that the criteria might be too strong to be necessary conditions on an arithmetization for it to be a contentful expression, among other reasons because they imply not only that ϕ is provable only if $Thm_T(\overline{\phi})$ also is, but also that, given the provability of ϕ, there exist numbers $\overline{\phi}$, $\overline{Thm_T(\overline{\phi})}$, $\overline{Thm_T(Thm_T(\overline{\phi}))}$, etc., *ad infinitum*. This latter claim, though, seems like an ontological assumption very far removed from the notion of ϕ's provability.[5] On the other hand, certain basic properties about provability do not appear in Feferman's list, and in fact cannot. Examples are reflection principles of the form "If ϕ is provable then it is true" which seem very central to the notion of provability but which (by Löb's theorem) are not provable for predicates for which Feferman's other criteria are provable.

Moreover, it is not clear that conditions of intensional adequacy should be the same for every mathematical system regardless of the strength of that system. One might expect, for example, that because of its limited resources a weak system would think of consistency in essentially a different way than PA does. But this only complicates matters further, for it seems unlikely that sets of conditions constitutive of the way T thinks about its own consistency should be determined in any uniform way. Thus Feferman's proposal is incomplete as a method for the explicit arithmetizations needed for a fully mathematical treatment of metatheory.

4.4 THE INSIDE/OUTSIDE DISTINCTION

A curious feature of Feferman's approach to intensionality is that he considers which formal representations of a theory's axioms give rise to the construction of intensionally correct predicates but does not consider the variety of ways to construct predicates once a representation of the theory's axioms is fixed.

Feferman actually discusses both sorts of consideration, but only to dismiss those of the second sort. Considerations of the appropriate manner to arithmetize a theory he describes as considerations "from the inside," while those about the arithmetization of

---

[5] Andrew Boucher *FOM Digest* Vol. 30 Issue 1, makes just this point.

theorem-hood and the like he calls considerations "from the outside." Calling a change in one's arithmetization a change "from the outside" is meant to connote artificiality, and Feferman dismisses these sorts of considerations as technical tricks that can be useful for solidifying formal results but not for generalizing metatheoretical findings. The reason, he says, is that changes from the outside essentially are "changes in the notion of logical derivation" ([1960], p. 39).

In the strong arithmetics Feferman studied, like PA, "changes in the notion of logical derivation" from standard provability are of two types. Either they are such radical changes as to be departures from any plausibly meaningful notion of provability, or the new notion is provably equivalent in the theory to the standard one. Therefore such changes are perhaps rightly dismissed as irrelevant for intensional purposes.

But weak arithmetics call for care in the arithmetization of syntactic properties like the notion of derivation, since only some of the many extensionally equivalent notions may be interpretable in the theory. Questions of intensionality accordingly extend in these theories to choices at every stage of arithmetization, from the representation of axiom-hood to more complex constructions of "metatheoretical predicates" on top of these representations.

For example, the cut rule

$$\frac{\Gamma \longrightarrow \Delta, A \qquad\qquad A, \Gamma \longrightarrow \Delta}{\Gamma \quad \longrightarrow \quad \Delta}$$

can be shown to be redundant in the standard sequence calculus presentation of first-order logic by a routine semantic argument, but this argument is only as good as the semantic theory on which it is based. One might argue that this theory is as good as one could ask for, but in so doing one would essentially dodge certain metatheoretical questions. For if one cannot execute the argument within an arithmetic theory, then the argument is useless towards gaining ground on questions of that theory's consistency.

Cut elimination can be formalized in PA. Thus in PA the formulas $\exists x\, Prf(x, y)$ and $\exists x\, CFPrf(x, y)$ expressing in turn "there is a proof of y" and "there is a cut-free proof of y" are equivalent. But if the *Hauptsatz* cannot be formalized in a weaker theory T, then from the standpoint of T the theorem says nothing about T-provability. In that case, only one of $\exists x\, Prf(x, y)$ and $\exists x\, CFPrf(x, y)$ can be intensionally correct. Feferman seems to think that in cases such as this the ordinary construction always will have more claim to correctness. Cut-free consistency and provability are, by comparison to the ordinary constructions, contrived. However, one wants to be able to distinguish the notions from within T, and there is no guarantee that what appears more artificial from a richer background will still appear so from T's standpoint. Indeed, the elimination of a deduction rule results in a syntactically simpler proof calculus, which T might more readily arithmetize in a meaningful way.

The inside/outside distinction is more illusory than it first appears, though, as a closer look at the relationships among arithmetical systems reveals. One of the weakest arithmetical theories is Robinson's Q (defined by Tarski, Mostowski, and Robinson in [1953]), which can be axiomatized in six axioms:

1. $\forall x\, S(x) \neq 0$
2. $\forall x \forall y (S(x) = S(y) \rightarrow x = y)$
3. $\forall x (x \neq 0 \rightarrow \exists y (S(y) = x))$
4. $\forall x (x + 0 = x)$
5. $\forall x \forall y (x + S(y) = S(x + y))$
6. $\forall x (x \cdot S(y) = x \cdot y + x)$

Extended conservatively with a definition for inequality

$$x \leq y \Leftrightarrow \exists z(x + z = y),$$

Q is the standard base subtheory of most arithmetic systems studied today. It is strengthened usually by the inclusion of an axiom schema or rule for some combinatorial principle, or by adding an axiom that stipulates that certain functions are total. For example PA is Q together with the rule

$$\frac{A(b), \Gamma \longrightarrow \Delta, A(b+1)}{A(0), \Gamma \longrightarrow \Delta, A(t)} \qquad \text{where the \emph{eigenvariable} } b \text{ does not occur except as indicated}$$

for induction on all formulas. Elementary Arithmetic $(I\Delta_0 + Exp)$ is Q with the induction rule restricted to bounded formulas and the axiom

$$Exp \Leftrightarrow \forall x \forall y \exists z(z = x^y)$$

saying that exponentiation is a total function.[6]

Since Q is finitely axiomatized, the encoding of Q-axiomhood can be done simply by listing Q's axioms. This arithmetization is optimal, in the sense that if $q(x)$ is the formula that defines Q simply by listing its axioms, then for any other definition $\kappa(x)$ of Q's axioms in Q, $Q \vdash Con_\kappa \rightarrow Con_q$. Feferman's considerations about how best to arithmetize axiom-hood thus arise only for theories that cannot be finitely axiomatized. But then the representability of these stronger theories as Q extended by rules in the sequent

---

[6] Perhaps the most well known fact about bounded arithmetic is that the exponential function is not definable in any bounded theory. This presents an obstacle even for writing the axiom Exp. The undefinability of exponentiation is due to Parikh [1971], and in the same report Parikh showed a way around this obstacle by explaining that the predicate $P(x, y, z) \Leftrightarrow x^y = z$ defining the graph of the exponential function is concrete. The graph of the exponential function is defined explicitly in bounded arithmetic in Hájek and Pudlák [1993], p. 301. Thus one may write the axiom Exp based on this predicate.

calculus[7] dissolves the "inside/outside" distinction. One can simply represent axiom-hood in terms of the efficient encoding of Q, and all remaining considerations will be in terms of constructing predicates from this encoding. The activity of choosing how to formulate provability and other metatheoretical notions then becomes the primary stage for questions of intensionality.

In turning now to a more general approach to intensionality, we thus return to considerations of how best to formulate provability and consistency informally, so that the encoding of these notions will be recognizable as such within the theory itself. This "consideration from the outside" has already been made in a most satisfactory way by Herbrand in his restatement of provability in terms more arithmetical, for it turns out that encoding his formulation of provability in T results in a formula immediately relativized to T's deductive strength. The result is a representation of T's metatheory within T that is accurate from the standpoint of T.

4.5   A THEORY-DEPENDENT INTERPRETATION OF CONSISTENCY

Kreisel [1958] argues that since the no-counterexample interpretation of provability is in a sense simple, it is a more primitive interpretation than the standard one and therefore is more appropriate when one restricts one's attention to constructive methods. He has in mind specifically recursive methods, but his sentiment can be extended to the kinds of restrictions one encounters when investigating arithmetization in weak theories. The no-counterexample

---

[7] In [1985] Sieg demonstrates that the presentation of arithmetic theories with induction as an inference rule is equivalent to the presentation with axiom schemata for induction. His demonstration shows that the induction rule is strong enough only with the inclusion of the side formulas A. For a detailed description of the proof-theoretical benefits of the rule-based presentation, his paper is an excellent resource, as is §1.4 of Buss [1998c].

interpretation is based on Herbrand's theorem in such a way that theories not proving Herbrand's theorem will not prove that an encoding of the no-counterexample formulation is equivalent to the standard encoding of theorem-hood. Kreisel's intuition, translated into this arena, is that this testifies against the meaningfulness of the standard formulation of theorem-hood for these theories. Since Herbrand proofs are propositional proofs, they are combinatorially very simple. One might say that weak theories are able to make sense of them when they are not able to make sense of the combinatorially more complex sequent or predicate calculus proofs. Herbrand proofs are at the same time generally much longer than standard proofs and therefore more difficult to execute. Kreisel's student Richard Statman [1978] proved in fact that no Kalmar elementary procedure transforms standard proofs into these combinatorially simpler "direct proofs." The intuition that seems appropriate is that meaningful results should be more difficult to prove than meaningless ones.

What concretely can be said about the comparative meaningfulness of standard formulations of consistency and the nonstandard one based on Herbrand's theorem? In weak theories where the two notions separate, it seems that there is quite a lot to say. The formula Con_T says that there are not proofs from the axioms of T of a sentence and of its negation. The formula $HCon_T$ says that there is not a propositional proof of the quantifier free matrix of a Herbrand disjunction:

$$\bigvee_{i=1}^{k} (He[\bigwedge_{j=1}^{l} T_j \to \perp](t_{i1}, f_1(t_{i1}), t_{i2}, f_2(t_{i1}, f_1(t_{i1}), t_{i2}), \ldots))$$

(4.1)

where $He[\phi]$ is the open part of the Herbrand form of ϕ, for some finite conjunction $\bigwedge_{j=1}^{l} T_j$ of T's axioms.

To analyze this construction, let some finite conjunction $\bigwedge_{j=1}^{l} T_j$ of T's axioms be given so that

$$\bigwedge_{j=1}^{l} T_j \iff \forall x_1 \exists y_1 \forall x_2 \exists y_2 \cdots \Phi(x_1, y_1, x_2, y_2, \ldots)$$

and consider the negation of the right-hand side of the above sentence:

$$\exists x_1 \forall y_1 \exists x_2 \forall y_2 \cdots \neg \Phi(x_1, y_1, x_2, y_2, \ldots).$$

A natural way to interpret formulas with alternating quantifiers is in terms of a two player "adversary game." In this case, the above formula says that there is a strategy for Eloise ("playing the existential quantifiers" against Abelard) to satisfy the open formula $\neg \Phi$. This is interpreted as Eloise demonstrating the inconsistency of T. According to Herbrand's theorem, the formula is provable in the predicate calculus just in case there is a propositional quasi-tautology of the form:

$$\bigvee_{i=1}^{l} \neg \Phi(t_{i1}, f_1(t_{i1}), t_{i2}, f_2(t_{i1}, f_1(t_{i1}), t_{i2}), \ldots)$$

where the t_{ij} are terms in the language of Φ expanded to include the function symbols f_j.

This gives the original formula (4.1) with Φ standing in place of the finite conjunction of T's axioms. One would like to recover the adversary game semantics for this object. Intuitively the functions f_j should compute Abelard's moves based on the previous moves, and the terms t_{ij} represent Eloise's moves. But on which previous moves are Abelard's moves based, and on what grounds does Eloise choose her moves?

According to the way terms and Herbrand functionals are introduced in the construction of the disjunction (see the restrictions on terms in the statement of Corollary 6 in Chapter 2), while Abelard's moves are defined by functions over shorter terms (earlier moves) in a single disjunct, Eloise's moves are not. *Her moves* in disjunct i may depend on what she knows of Abelard's move-computing functions from other disjuncts, because nothing prevents terms computed by functions from disjuncts i from appearing as subterms of terms t_{jk} from other disjuncts. That is, once a function symbol has been introduced into the language (by a play of Abelard), Eloise may use it in the terms that make up her future plays in *any* of the disjuncts. So in the adversary game, the individual disjuncts are constructed in parallel rather than sequentially. In typical cases in fact the parallel construction is necessary in order to arrive at the disjunction guaranteed by Herbrand's theorem.[8] For example given a formula whose prenex-normal form has the quantifier prefix $\forall\exists\forall$, one can write down the general form of a Herbrand disjunction as

$$\bigvee_{i=1}^{l} \phi(z_0, t_i(z_0, z_1, \ldots, z_{i-1}), z_i), \tag{4.2}$$

and for the quantifier prefix $\forall\exists\forall\exists$, the general form can be written

$$\bigvee_{i=1}^{l}\bigvee_{j=1}^{k_i} \phi(z_0, t_i(z_0, z_1, \ldots, z_{i-1}), z_i, s_{ij}(z_0, z_1, \ldots, z_l)), \tag{4.3}$$

[8] This interpretation of Herbrand disjunctions is essentially due to Adamowicz and Kolodziejczyk [2004]. See there or Pudlák [2004] for the necessity of the parallel construction. Pudlák in fact presents Adamowicz and Kolodziejczyk's analysis in terms of games, similar to the presentation in this study. They use this analysis to define combinatorial principles independent of bounded arithmetic but do not discuss questions of intensionality. Pudlák, interestingly, does emphasize that although the combinatorial principles he treats are essentially Π_2 and consequently in a mathematical sense less interesting than other, well-known independent sentences, the game semantics suggests that they are more meaningful in weak settings than the familiar, arithmetically less-complex sentences.

but for formulas whose prenex-normal forms begin with more complex quantifier prefixes, general forms cannot be written. This is because in the case of (4.2) Eloise cannot do better than to base her move in disjunct i on what she knows of Abelard's replies to her moves in all previous disjuncts. Similarly, in the case of (4.3) Eloise would hurt her own chances of constructing a counterexample if she did not delay her second move in each disjunct until she knew how Abelard had replied to her first move in each disjunct. But in more complex cases, Eloise must always make choices about whether to maximize her information before proceeding to any of her nth moves, or to make some nth moves in some disjuncts and observe how Abelard replies to them before making some of her $(n - 1)$th moves in other disjuncts.

The semantic interpretation that arises is this. Abelard and Eloise play the game Φ given by the formula $\exists x_1 \forall y_1 \exists x_2 \forall y_2 \cdots \neg \Phi$ $(x_1, y_1, x_2, y_2, \ldots)$ on several boards (indexed by i), with Abelard boasting a winning strategy for Φ (given by the functions f_j). Eloise's goal is just to win on at least one board, thereby disproving Abelard's boast. If Eloise succeeds, then she will have proven T inconsistent. In that case the terms she plays define a winning "superstrategy" – that is, these terms define a way to reply to Abelard's alleged strategy for the game Φ either by replying directly to a move of Abelard on the board where he just played, or by using the information learned about a position from one of Abelard's moves by making a different move on another board, or by beginning a game on a new board, adding thereby to the number of disjuncts.[9]

---

[9] The scenario is analogous to a human player (Eloise) playing chess against a chess program (Abelard) with an option to take back her moves. The computer claims to have a strategy making it the Chess Master© and always plays this strategy. The human player, on the other hand, can at any point rethink a particular move, return to that stage of the

A tautological Herbrand disjunction of this form would be "a counterexample to T." We may formalize the claim that there is no counterexample as

$$\forall \overline{\Delta}(HD(\overline{\neg \Phi}, \overline{\Delta}) \rightarrow Tr_\exists(\overline{\exists z_1 \cdots \exists z_m \neg \Delta(z_1, \ldots, z_m)}))$$

where $Tr_\exists(x)$ is a truth definition for existential formulas and $HD(x, y)$ is the relation that says that y is a Herbrand disjunction for x. In our informal semantics, this last formula says that Abelard has a winning strategy for any Herbrand disjunction that would be, if Eloise had a winning superstrategy, a counterexample to T's consistency.

It is tempting to think in this case of Herbrand's theorem as showing us that the claim that no such formula $\Delta(z_1, \ldots, z_m)$ is a tautology is a claim about the consistency of T. Herbrand, as discussed in Chapter 3, thought rather that the consistency question had not been phrased precisely until with his Fundamental Theorem he reduced it to arithmetical terms. Thus when he studied Gödel's results in 1931 he was somewhat hesitant about the intensional correctness of the second theorem. In the [1931d] paper he submitted to the *Journal für die reine und angewandte Mathematik* just before his fatal hiking trip, he summarizes Gödel's theorem thus: "The consistency of a theory cannot be proved by arguments formalizable in the theory, whenever the theory contains arithmetic." This is, of course, the informal paraphrase of the theorem that Gödel used, which is accurate only if his arithmetization is intensionally correct. Herbrand seems to have been aware of the difficulty in arguing

game, and substitute a different move, all the while keeping the board with the original move on it "alive" in case later it appears preferable after all to play from that position. Since human time-resources are valuable, most chess players are aware of how difficult it is to beat good programs even with the option to take back moves. On the other hand, since we do not have access to the best strategies due to the combinatorial explosion in chess positions, an option to take back moves increases our chances substantially.

for the intensional correctness of Gödel's formulas for theorem-hood and consistency, as his gloss to his summary of Gödel's second theorem suggests:

To understand this proposition, one has to imagine that all signs occurring in metamathematics are represented by objects of the theory being considered, for example, by integers in arithmetic; the properties of these signs and the relations among them will then be represented by certain propositions of the theory; every argument in the theory in question carried out with these objects and *these propositions will correspond to a metamathematical argument, of which it will be, in a way, the translation.* ([1931d], p. 295, emphasis added)

Thus it seems that since Gödel's encoding of metatheory in arithmetic bypasses Herbrand's reformulation of meta-mathematical questions in arithmetical terms, Herbrand was forced to qualify the degree to which Gödel's predicates accurately express the metatheoretical notions they are meant to translate. Following Herbrand, then, I want to suggest that the provability of Herbrand's theorem in a system would invest the standard consistency statement with meaning, by virtue of the meaning of the claim that no such Herbrand disjunction exists, but that the theorem's failure in a system casts doubt on the meaningfulness in that setting of the standard consistency claim. In this case the "no-counterexample" construction is the only one interpretable in T as a consistency statement. To establish this concretely, one must first see why the encodings of Herbrand's formulations of metatheoretical claims do express those claims in a way recoverable by the theory in which they are encoded. The reason, which is presented below, is that the no-counterexample interpretation of provability can be relativized to individual arithmetical systems.

What is especially useful about Herbrand's formulation of consistency is that the informal interpretation in terms of match strategies can be adapted to the computational strength of any arithmetical system. Consider the formula embedded in the consequent of the above "no-counterexample" claim:

$$\exists z_1 \cdots \exists z_m \neg \Delta(z_1, \ldots, z_m).$$

This formula says not only that $\Delta(z_1, \ldots, z_m)$ is not a tautology but also that Abelard can substitute values for the variables z_i in such a way that he beats Eloise's strategy (defined by the terms t_{ij} in Δ) "on every board." Extensionally, this sentence is stronger than the mere claim that $\Delta(z_1, \ldots, z_m)$ is not a tautology, because its truth depends on what substitutions are effective in T (the substitution functions must be provably total in T). Likewise $\bigvee_{i=1}^{l} \neg \Phi(t_{i1}, f_1(t_{i1}), t_{i2}, f_2(t_{i1}, f_1(t_{i1}), t_{i2}), \ldots)$ is stronger than a mere claim of inconsistency, because Eloise's counterexample construction strategy must be computable by terms t_{ij} of the language of T. In all bounded theories, however, these strengthenings are unavoidable in meaningful talk about strategies, for if moves require for their determination the results from a function that one cannot even prove is total on the natural numbers, then this requirement undercuts the effectiveness of a proposed strategy. One has in that case no strategy at all.

In particular, in bounded theories like $I\Delta_0 + \Omega_1$ or Buss' S_2^1, one cannot define the terms needed to construct Herbrand disjunctions for provable formulas, because the terms needed are values of functions that those theories cannot prove total. The failure of Herbrand's theorem for these theories in fact amounts just to this. The tautological Herbrand disjunction one gets from the theorem applied to provable formulas often involves $I\Delta_0 + \Omega_1$- or S_2^1- uncomputable terms. In particular the consistency statements for these theories are equivalent, not to the claim that Abelard can beat any superstrategy that Eloise plays, but to the non-existence of tautological Herbrand disjunctions that are not meaningfully interpretable. One can then bifurcate the notion of "Herbrand consistency" into two distinct claims, neither provably equivalent in bounded arithmetic to the standard consistency statement. For the first the existence of all possible counterexample disjunctions is considered. Though not T-provably equivalent to the consistency statement expressed in terms of derivability in the sequent calculus, a Gödel-like theorem is provable for this statement.[10] A second statement allows in the disjunctions only terms whose values are polynomial in the size of their subterms, so that the existence of a counterexample depends on plays calculable with the resources of T. However, it is not known how to prove a Gödel-like theorem with this restriction or whether such a theorem can be proved.[11]

Kreisel's intuition can now be reformulated in terms of these constructions. The straightforward two-player game semantics for the predicate calculus can be adapted to a weak notion of provability based on the construction of Herbrand disjunctions. Game strategies become match strategies, where the same game is played on several boards. Abelard plays on all boards the unique move that he deems best in that position, while Eloise is free to play differently on every board. (This restriction on Abelard and license for Eloise is the exact reverse of the scenario from the standard quantifier semantics. This results from Eloise trying to falsify the originally quantified predicate of interest rather than trying to

---

[10] ...in many weak arithmetics. Willard [2002] discusses how strong a theory one needs in order to prove Gödel-like theorems for direct proof systems. Among his results is a proof that $I\Delta_0$ suffices, as conjectured by Wilkie and Paris [1981].

[11] Krajíček and Takeuti [1992] study a notion of restricted provability, which requires that all terms be provably recursive, and show that the second incompleteness theorem fails for this notion.

confirm it.) Both players' search strategies, however, are bounded by the terms of the theory. From the "point of view" of the theory under consideration, though, this restriction is perfectly natural. What results is not a privation of strategies for the game at hand, only a realistic focus on strategies that one actually can find (as opposed to a hypothesized realm of ideal strategies, the existence of which is more dubious than the consistency of the theory in question).

The no-counterexample interpretation applied to the question of consistency, then, results in a semantics relativized to the computational strength of the theory in question. The same cannot be said about the standard formulation of consistency, which relies still on the arithmetization of combinatorially complex proof calculi. The significance of the respective ability and inability of these two formulations of consistency to relativize is evident in weak theories that do not prove those formulations' equivalence. The intensional correctness of formulations based on the no-counterexample interpretation together with the unprovability in T of its equivalence with the standard construction draws into question the meaning of meta-mathematical results based on the standard construction. In fact in such a weak setting one may not be able sensibly to speak about the possibility of deriving a contradiction, because that claim is equivalent to a claim about "ideal" match strategies that are not computable in the system.

4.6 CONCLUSION

This suggests a limitation to the pursuit of intensionality in the way Feferman proposes. His criteria express facts that at first sight seem closely tied in to the notion of provability, but since sufficiently weak theories do not prove that proof by natural deduction is equivalent to proof *via* construction of Herbrand disjunctions, the evident interpretability of the latter (in theories strong enough to interpret *it*) suggests that arithmetization schemes satisfying the "intensionality criteria" need not capture adequately the notions of provability and consistency after all. Since in theories of bounded arithmetic, the standard consistency statement is provably equivalent to the theory's Gödel sentence, but the Herbrand consistency statement is not, "the consistency of bounded arithmetic," meaningfully construed, is not equivalent to such theories' Gödel sentences.

Instead of proposing such criteria, one can begin with a formulation of consistency that, once encoded directly as a formula of T, is already relativized to the strength of T. Such a formulation is the one Herbrand produced with his Fundamental Theorem. The failure of conditions like the ones Feferman proposes for these formulas then appears like evidence of the inadequacy of Feferman's conditions rather than like evidence of the intensional incorrectness of these formulas.

In this way a combination of Herbrand's considerations "from the outside" (about how to formulate metatheoretical properties in arithmetical terms) with Gödel's techniques (for encoding such formulations as actual sentences of arithmetic) yields a treatment of metatheory with a strong claim to being purely mathematical. In particular, the treatment always will be sensitive to the particular theory under investigation, in keeping with the general conception of mathematical purity in metatheoretical studies that Hilbert as early as in *Grundlagen der Geometrie*. It is an interesting question whether the results obtained when one pursues metatheoretical questions in this way differ from the traditional results. In the next chapter I apply this outlook to the case of Gödel's second incompleteness theorem for Robinson's Q and find that the results can differ quite drastically. This indicates that a programmatic study of meta-mathematics according to Hilbert's standard of purity should diverge from the results of meta-mathematical investigations according to classical techniques.

CHAPTER 5

Interpreting Gödel's second
incompleteness theorem for Q

In the 1950s Georg Kreisel posed the question whether the provability of Gödel's second incompleteness theorem for Robinson's arithmetic Q, if such a proof were ever discovered, could be seen as a demonstration of "the unprovability in Q of Q's consistency." According to Kreisel this question arises because Gödel's techniques were not adequately arithmetical: "Gödel's work on formulae expressing the consistency of classical arithmetic goes beyond arithmetic concepts because it uses metamathematical interpretation" ([1958], p. 177). In Gödel's [1931] paper, arithmetical formulas "express" metatheoretical properties like consistency through binumeration. Since a binumeration is a correlation between some formal sentences and some yet-to-be-formalized meta-mathematics, the binumeration cannot be verified by purely arithmetic means. That is, the *binumeration* cannot be arithmetized. Thus in order to verify that a formula expresses consistency, one must step outside of the arithmetical setting. Gödel had not shown that an arithmetic theory could pose the question of its own consistency on its own terms. Kreisel suggested that this might be possible, but that in very weak settings like Robinson's theory it was unlikely.

In [1960] Solomon Feferman proposed a method to allow arithmetical theories to formulate statements about their own metatheory more directly. He presented a style of arithmetization in which one need not rely on evaluation outside of the setting of arithmetic to determine whether a formula properly expresses a meta-mathematical property. The technique proceeds thus: One wants a sentence of arithmetic to express, for example, that arithmetic is consistent. First one characterizes what sorts of mathematical properties any claim of consistency must have. Then one proves within arithmetic that a certain sentence has these properties. Now, whether or not this sentence binumerates the consistency claim, the arithmetic one is using proves all the salient properties one needs to know about consistency. Therefore the arithmetic itself "thinks" that the sentence expresses consistency.

Feferman calls his arithmetical techniques "intensional" because they take place within the mathematical setting one is studying, as opposed to Gödel's original "extensional" techniques which can be verified as accurate only outside of that setting. In the vocabulary Feferman introduced, Kreisel's intuition can be formulated more precisely: Gödel's second incompleteness theorem cannot be proved for Q with an intensional arithmetization.

Whether Gödel's second theorem could be proved for Q through *any* means remained an open problem until Bezboruah and Shepherdson's [1976] proof. As Kreisel predicted, their proof was essentially extensional: Not only did they not use Feferman's explicit arithmetization techniques, but the standard properties of provability given by the Hilbert–Bernays–Löb derivability conditions were not provable in Q for their provability predicate. Thus their proof did not resemble Gödel's original proof but required an essentially new idea. This led them to caution against trying to understand their theorem as the unprovability in Q of Q's consistency:

We must agree with Kreisel that this [result] is devoid of any philosophical interest and that in such a weak system this formula [Con_Q] cannot be said to express consistency but only an algebraic property which in a stronger system (e.g., Peano arithmetic P) could reasonably be said to express the consistency of Q. (p. 504)

Recently Pavel Pudlák has argued that a version of Gödel's second theorem for Q *can* be seen as a demonstration of the unprovability in Q of Q's consistency. Pudlák's argument is based on a strengthening of Bezboruah and Shepherdson's theorem. According to his result, Q proves that it is consistent to assume that there is a contradiction encoded already by a number in any proper syntactic cut of its numbers. Such a syntactic cut can be shortened using a method of Solovay so that it is a definition in Q of a model of a stronger arithmetic theory ($I\Delta_0 + \Omega_1$) that is supposedly able to express consistency unproblematically. Pudlák argues that this allows a sort of semantic bootstrapping from Q to $I\Delta_0 + \Omega_1$. Reworded in Feferman's terms, Pudlák's claim is that an intensional arithmetization of Q's metatheory is possible in $I\Delta_0 + \Omega_1$, and that Q can borrow this arithmetization through an interpretation of $I\Delta_0 + \Omega_1$. The provability of a strong version of Gödel's second theorem *via* this interpretation then refutes Kreisel's intuition by providing an intensional version of the theorem for Q. Pudlák's argument is presented in Section 5.2.

In this chapter I defend Kreisel's intuition. I argue contrary to Pudlák, first that in theories of bounded arithmetic like $I\Delta_0 + \Omega_1$ it is not fully evident how to develop an intensional arithmetization of metatheory. Then I argue that even if an intensional arithmetization of Q's metatheory were possible in $I\Delta_0 + \Omega_1$, the interpretability of this theory in Q would not result in an intensional arithmetization in Q itself.

Since either argument suffices to refute Pudlák's claim, let me explain this double-barrelled strategy. The first point is important *vis-à-vis* Hilbert's insistence that metatheoretical questions be investigated purely mathematically. Intensionality allows one to avoid semantic assumptions in one's treatment of metatheory, and so a Hilbertian investigation of metatheory for weak arithmetics should be sensitive to the expressive power of the system one studies. Section 5.3 contains a brief development of a formulation of consistency based on Herbrand's theorem that arguably has more claim to being an "intensional expression" of consistency in weak fragments of arithmetic than the standard formulation. The second observation discloses a mistake in reading too much philosophical significance into the technique called theoretical interpretation. Precaution on this matter seems due since interpretability results often are cited as significant for projects in the foundations of mathematics. Section 5.4 is a detailed demonstration of the inability of transferring the semantic expressability of $I\Delta_0 + \Omega_1$ into Q through interpretation.

5.2 PUDLÁK'S ARGUMENT

In [1996] Pudlák writes:

Bezboruah and Shepherdson proved the second incompleteness theorem in Q, which is one of the weakest arithmetic theories. The result was in a certain sense problematic: if the theory is so weak, does the particular formulation of Con_Q really mean what was intended? A solution ... is to define an initial segment J of the numbers in Q, which is an inner model of a stronger theory T and prove that it is consistent to assume that a proof of contradiction from axioms of Q is encoded by a number which is already in J. Since T is strong, the meaning of Con_Q is not so ambiguous [in T]. (p. 66)

After presenting the proof that it is consistent with the axioms of any theory containing Q to assume that there is a proof of contradiction from those axioms encoded in any inductive cut of the theory's numbers, he adds: "A corollary of this result [from Pudlák [1985]] is that the second Gödel's incompleteness theorem holds in weak theories, in particular in Q, without any doubts about what Con_T really means there. This is because by [Wilkie's interpretation of $I\Delta_0 + \Omega_1$ in Q] there is a cut in Q which is a model of $I\Delta_0 + \Omega_1$[1]. In such a cut all reasonable definitions of Con_Q are equivalent" ([1996], p. 71).

The claim is that since

1. in the theory $I\Delta_0 + \Omega_1$ the consistency of Q is unambiguous,
2. this theory does not prove the consistency of Q, and
3. $I\Delta_0 + \Omega_1$ is interpretable in Q,

one is able to establish not only Gödel's second incompleteness theorem for Q but also "the unprovability of Q's consistency." In the introductory section we described this argument as a bootstrapping argument. Pudlák implicitly concedes that Bezboruah and Shepherdson's proof of Gödel's theorem in Q is not metamathematically meaningful, that Robinson's theory is too semantically impoverished to recover the meaning of Con_Q intensionally. But he claims that in $I\Delta_0 + \Omega_1$ an intensional arithmetization of Q's metatheory is possible. That is, $I\Delta_0 + \Omega_1$ proves facts about the predicate Thm_Q that suffice for it to be an actual statement of provability in Q. Pudlák suggests that Q can in a

---

[1] $I\Delta_0 + \Omega_1$ is a common theory of bounded arithmetic. Theories of bounded arithmetic are extensions of Parikh's [1971] PB, which is Q together with an induction schema restricted to bounded formulas. $I\Delta_0 + \Omega_1$ extends PB by including the axiom $\Omega_1 \Leftrightarrow \forall x \forall y \exists z (z = x^{(\log y)})$ saying that the function $\omega_1(x, y)$ with polynomial growth rate is total. The function $\omega_1(x, y)$ is useful for sequence coding.

sense borrow this stronger theory's semantic richness by proving theorems whose quantifiers all are relativized to a formula that models the stronger theory. Ordinary bootstrapping is the emulation of certain purely formal constructions in a weak theory through such relativizing of quantifiers or conservative extensions by definition. The ability of Q to do this is uncontroversial: By relativizing the predicate Thm_Q to an interpreting domain for $I\Delta_0 + \Omega_1$, one generates a formula for which Q proves the derivability conditions as well as other properties that $I\Delta_0 + \Omega_1$ proves about Thm_Q. However, Pudlák's claim is stronger. His point is that not only are these properties provable for this formula, but that they also characterize intensional correctness even in Q.

Pudlák's argument fails on two points. First, the "strong" theory $I\Delta_0 + \Omega_1$ does not treat consistency so unambiguously. There are several extensionally equivalent formulations of consistency that are not provably equivalent in $I\Delta_0 + \Omega_1$, and this theory does not prove the unprovability of each of them. Second, formal interpretability fails to recover the intensionality of an arithmetization. The intensional arithmetization of syntax is meant to invest the meta-mathematical results following out of this arithmetization with a secure meaning by eliminating the need to step out of the framework of a theory in order to verify what its formulas express. But if the meaning of a sentence about the consistency of one theory depends on an interpretation in that theory of a stronger theory and the properties that *it* proves about that sentence, then there is no epistemological gain. This is because the adequacy of intensionality criteria is theory-dependent: Even if some criteria are recoverable from a strong theory through its interpretation, those critera need not be adequate for intensionality in the weaker, interpreting theory.

5.3   TWO ROUTES TO INTENSIONALITY

The first failure in Pudlák's argument has to do with determining when a set of criteria adequately capture an intended metatheoretical notion. Even in theories as strong as PA, not everything true and important about theorem-hood is provable about the predicate Thm_{PA}. For example, Gentzen's consistency proof and the classical form of Gödel's second theorem together show that PA does not prove cut-elimination up to the ordinal ϵ_0. Feferman's method thus assumes the possibility of isolating criteria that sufficiently characterize theorem-hood – the idea being that cut-elimination up to the ordinal ϵ_0 is only an incidental, not an essential, property of proofs. The prospects for this isolation appear dim, however, when one recalls that some of the most intuitive facts about provability in a theory are demonstrably not provable in the theory. Consider an illustrative example: One property that one might expect from an intensional arithmetization of theorem-hood in PA is the reflection principle

$$Thm_{PA}(\overline{\phi}) \rightarrow \phi,$$

for all sentences ϕ. However according to the formalized Löb's theorem,

$$PA \vdash Thm_{PA}(\overline{Thm_{PA}(\overline{\phi}) \rightarrow \phi}) \rightarrow Thm_{PA}(\overline{\phi}),$$

reflection can only ever be proved for provable formulas. Concerning PA's surprising "admission" that it does not prove that the predicate Thm_{PA} is reflective for any unprovable formulas, Rohit Parikh has joked that "PA couldn't be more modest about its own veracity." The question that one faces when pursuing Feferman's route to intensionality is whether this much modesty is consistent with a sufficiently robust notion of theorem-hood.

Notwithstanding this evident difficulty, Feferman's suggestion that such an isolation is possible has been influential. In [1998c] Buss follows Feferman in suggesting that the theorems listed here suffice for an intensional arithmetization in T:

$$T \;\vdash\; \forall u \forall v \forall w (Fmla(u) \wedge Term(v) \rightarrow Fmla(\overline{sub(u, v)}))$$
$$T \;\vdash\; \forall \overline{\phi} \forall \overline{\psi} (Thm_\tau(\overline{\phi}) \wedge Thm_\tau(\overline{\phi \rightarrow \psi}) \rightarrow Thm_\tau(\overline{\psi}))$$
$$T \;\vdash\; \forall u (Proof_\tau(u) \rightarrow Thm_\tau(\overline{Proof_\tau(u)}))$$
$$T \;\vdash\; \forall \overline{\phi} \forall u (Prf_\tau(u, \overline{\phi}) \rightarrow Thm_\tau(\overline{Prf(u, \overline{\phi})}))$$
$$T \;\vdash\; \forall \overline{\phi} (Thm_\tau(\overline{\phi}) \rightarrow Thm_\tau(\overline{Thm_\tau(\overline{\phi})}))$$

Although Feferman does not stipulate any list of criteria as necessary or sufficient for an intensional development of arithmetization, we shall refer to these as the Feferman criteria. (Notice that the Feferman criteria imply the Hilbert–Bernays–Löb derivability conditions.) This book contains no consideration of the question whether Feferman's account of intensional correctness is satisfactory for strong arithmetic systems like PA other than the general remarks above. It is worth mention, though, that there is no consensus about what might be proper conditions for intensionality even in such strong settings.[2] We here suggest only that deciding what the proper conditions are for the intensionality of an arithmetization itself depends on the strength of the theory one is studying. Thus even granting the propriety of the Feferman criteria for PA, more must be said about intensionality in systems like $I \Delta_0 + \Omega_1$.

To this end consider the formalization of theorem-hood based on Herband's [1931a] Fundamental Theorem. Herbrand's theorem allows a type of reduction of first-order provability to propositional provability. Since Herbrand proofs are propositional proofs, they

---

[2] In [1986] Detlefsen argues that Hilbert's foundational program, because of its underlying "instrumentalist" epistemology, is in theory realizable by proving in strong arithmetics formulas that fail to satisfy the derivability conditions.

are combinatorially very simple. It is therefore reasonable to think that weak theories might be able to make sense of them when they are not able to make sense of their associated, combinatorially more complex, quantificational proofs. Herbrand proofs are at the same time generally much longer than standard proofs, and therefore more difficult to execute. Statman [1978] proved in fact that no Kalmar elementary procedure transforms standard proofs into these combinatorially simpler "direct proofs." Therefore arithmetics that do not treat functions with super-exponential growth rate do not prove the Feferman criteria for these proofs – specifically they do not prove their closure under *modus ponens*. The intuition that seems appropriate is that meaningful results should be more difficult to prove than meaningless ones.

Let us investigate this intuition by considering a consistency statement based on the construction of Herbrand proofs. The formula $HPrf_T(\phi)$, read "there is a Herbrand proof of ϕ," says that for some finite set $\{T_j\}_{j=1}^{l}$ of the axioms of T there is a disjunction

$$\bigvee_{i=1}^{k}(He^*[\bigwedge_{j=1}^{l}T_j \to \phi](t_{i1}, f_1(t_{i1}), t_{i2}, f_2(t_{i1}, f_1(t_{i1}), t_{i2}), \ldots))$$

that is a propositional tautology ($He^*[\phi]$ is the quantifier-free part of the Herbrand form of ϕ). Herbrand's theorem states that a such a tautological disjunction exists if, and only if, ϕ is provable in the predicate calculus. The *Herbrand consistency* of T is thus the claim that for no finite set $\{T_j\}_{j=1}^{l}$ of the axioms of T is there a propositional proof of a disjunction of the form:

$$\bigvee_{i=1}^{k}(He^*[\bigwedge_{j=1}^{l}T_j \to \bot](t_{i1}, f_1(t_{i1}), t_{i2}, f_2(t_{i1}, f_1(t_{i1}), t_{i2}), \ldots)).$$

This sentence is abbreviated $HCon_T$.

The arithmetization of Herbrand consistency involves two steps. First, as usual, one must arithmetize the notion of axiom-hood in order to say "for no finite set of the axioms of T ..." Second one must arithmetize the predicate "x is a proof." The principal difference between this and the familiar arithmetization of consistency is that one is interested only in propositional proofs. Notice also that in the statement of Q's Herbrand consistency (or standard statement of consistency), there is no decision to be made about how to arithmetize axiom-hood: Since Q is finitely axiomatized, the direct coding of a list of the axioms of Q is maximally efficient.[3]

As noted above, Herbrand proofs are long. The tautological disjunctions thus proved also can be quite long. For this reason Herbrand forms of first-order formulas are customarily called "expansions" of these formulas. This is an ironic moniker for what Herbrand referred to as the *"réduite"* of a quantified formula. Despite their unwieldiness as objects, the propositional disjunctions guaranteed by Herbrand's theorem are *reductions* of quantified formulas because they are their extensional equivalents but are "constructive" or "finitary" due to the role that the Herbrand functions play as quantifier eliminators. They reduce semantically complex, quantified expressions to truth functional compounds of numerical terms.

This feature of Herbrand's theorem indicates a second route to intensionality. To say that there is a Herbrand proof of a formula ϕ is to say that there is no substitution of terms in the quantifier-free matrix of ϕ that falsifies this matrix in every disjunct of the *réduite*. Intuitively one thinks of this as a "no-counterexample

---

[3] If we let $Q(x)$ be the formula that defines Q simply by listing its axioms, then for any other definition $Q'(x)$ of Q, $Q \vdash Con_{Q'} \rightarrow Con_Q$ and $Q \vdash HCon_{Q'} \rightarrow HCon_Q$. This type of maximality is unattainable in general in theories that cannot be finitely axiomatized, so additional intensionality considerations arise. See Section 4.4.

claim." Herbrand's theorem thus provides what Kreisel calls a "no-counterexample interpretation" of first-order formulas: Given a theorem of a first-order theory T, Herbrand's theorem returns an equivalent, but essentially constructive, propositional tautology. Therefore if T proves Herbrand's theorem, then the statement in T that ϕ is a theorem of S is equivalent in T to the claim that S cannot construct a counterexample of ϕ. In the statement of the last sentence, we do not assume that "the statement in T that ϕ is a theorem of S" is meaningful in T, only that it is extensionally correct. However, since T equivocates this statement with a no-counterexample claim, and *this* claim is combinatorially simple, one may *infer* that T attributes the intended meaning to the original formula $Thm_S(\phi)$.

This route to intensionality is open in PA. Let us apply it to the case of the formula Con_Q. Since PA proves Herbrand's theorem, it also proves the biconditional $Con_Q \leftrightarrow HCon_Q$. Thus there is little reason to doubt that Con_Q is intensionally correct in PA: The formula $HCon_Q$ encodes only constructive metatheoretical notions, truth-functional connectives, recursive functions, and a finitary representation of Q. It is therefore inherently arithmetical in a way that Con_Q is not. But since the formulas are provably equivalent in PA, they "express" one and the same arithmetical, no-counterexample claim.

By contrast, Herbrand's theorem is *not* provable in $I\Delta_0 + \Omega_1$. This theory consequently cannot interpret the formula Con_Q *via* Kreisel's no-counterexample interpretation. A no-counterexample claim might in this theory be an intensionally correct expression of Q's consistency without its correctness translating to Con_Q, but the unprovability of Herbrand's theorem changes what that no-counterexample claim is, in a way that analysis of another example will illuminate.

Consider the formulas Con_δ and $HCon_\delta$, where δ is a recursively enumerable arithmetization of the axioms of $I\Delta_0 + \Omega_1$. The fact that $I\Delta_0 + \Omega_1 \nvdash Con_\delta \leftrightarrow HCon_\delta$ provides an analysis of Gödel's second theorem for this theory. When a theory proves the equivalence of a first-order formula and a Herbrand *réduite*, we know that the quantifier-eliminating functions in the *réduite* are provably recursive in that theory. As Kreisel observes,

the same analysis explains why ... certain true formulae $\forall x \exists y A(x, y)$ *cannot* be proved in a given system: namely, though they are true, i.e. though there is a (recursive) function f which satisfies $\forall x A(x, f(x))$, f does not belong to the class of functions which can be *proved* to be recursive in this system. So, if no function f which belongs to this class satisfies $\forall x A(x, f(x))$, and if $\forall x \exists y A(x, y)$ is true, then it is independent in the system considered. ([1958], p. 158)

Thus the unprovability of the standard consistency statement in bounded arithmetic appears merely to be a consequence of the fact that there are function symbols in these theories' languages that they do not prove to be functions. The "constructive content" of the formula Con_δ is not sufficiently constructive to be meaningfully interpreted by $I\Delta_0 + \Omega_1$. Recall that the Herbrand *réduite* of a formula expresses that there is no substitution of terms which refutes that formula. For an intensional recovery of this claim, the notion of substitution must be interpretable as such by the theory in question. Since the Herbrand *réduite* of a formula of bounded arithmetic might involve terms built up from functions that are not provably total, this object is not interpretable as a no-counterexample claim in the theory itself. One can recover intensionality by reformulating Herbrand provability with the restriction that only terms built up from $I\Delta_0 + \Omega_1$-provably total functions are used in constructing a formula's *réduite*, but it is not known whether this modified no-counterexample claim is provable.

A construal of no-counterexample claims as intensionally correct has the plausible consequence that the formulations of such metatheoretical properties as provablility and consistency depend on the computational strength of the theory in question: The "meaningful" expression of consistency in the impoverished setting of bounded arithmetic differs appropriately from PA's expression of its own consistency, not only in that a different theory is mentioned, but in that the very notion of what it would mean to be inconsistent has been reworded in the elementary terms with which the weaker theory is familiar. Perhaps most welcome in this development is that no decisions have to be made about which properties are essential to, or constituitive of, the intended metatheoretical notions and which are merely incidental. One rather can discover what a particular arithmetic "thinks" are the essential properties of proofs by invesitgating what it proves about the predicate $HPrf_T(x)$.

Rather than press this analysis further, it suffices to emphasize that while both routes to intensionality have their attractions, they are incompatible in weak fragments of arithmetic and specifically in the theory $I\Delta_0 + \Omega_1$. If one takes no-counterexample claims as meaningful expressions within a theory, then one must give up on the Feferman criteria in sufficiently weak settings. Alternatively, an endorsement of the Feferman criteria as constituitive of intensionality in bounded arithmetic involves rejecting the meaningfulness of no-counterexample claims – a penalty one need not pay in stronger settings. Thus we cannot agree that "all reasonable definitions of consistency are equivalent" in $I\Delta_0 + \Omega_1$. Meaningfully encoding the statement of Q's consistency in this setting turns out to be not so straightforward after all.

5.4   AMBIGUITY IN INTERPRETATION

Let us turn to the second failure in Pudlák's argument.

The interpretability of $I\Delta_0 + \Omega_1$ in Q alone does not recover the "content" of the former theory's theorems or the intensionality of its arithmetization for Q. If Pudlák's argument were simply this, it would apply also in familiar settings with shocking results. For example if we add to PA an axiom $\neg Con_{PA}$ saying that PA is inconsistent, the resulting theory is interpretable in PA.[4] By the "simple" argument we are considering, one would conclude that PA proves its own inconsistency!

Thus Pudlák's claim must not be simply a heuristic point about the significance of formal interpretability. One can do it more justice by understanding it rather as a prescriptive point about how to construct a predicate that unlike Con_Q is an intensionally adequate representation of provability in Q. The prescription depends not only on the interpretability of $I\Delta_0 + \Omega_1$ in Q but on a specific method of interpretation.

The Nelson–Wilkie interpretation of $I\Delta_0 + \Omega_1$ in Q is *via* a proper syntactic cut – a formula that Q proves is closed under successor and addition, but not induction. Formally a syntactic cut of Q's numbers is a formula $J(x)$ such that

$$Q \vdash J(0)$$
$$Q \vdash \forall x(J(x) \rightarrow J(x+1))$$
$$Q \vdash \forall x \forall y(y < x \wedge J(x) \rightarrow J(y)).$$

If additionally $Q \nvdash \forall x J(x)$, then J is a *proper* cut. PA, since it has a rule for induction on all formulas, does not have any proper syntactic cuts, but fragments of PA with induction restricted to

---

[4] This follows immediately from Theorem 6.6 of Feferman [1960].

formulas beneath a certain quantifier complexity may have proper syntactic cuts.

In fact, proper cuts arise as early in the arithmetic hierarchy as possible: To see that $I\Sigma_n$ has a proper cut of complexity Σ_{n+1}, let $n \geq 0$ and let $\Phi \Leftrightarrow \exists y \Psi(x, y)$ be any Σ_{n+1} formula for which $I\Sigma_n$ does not prove an induction instance (there must be some such Φ by the propriety of the arithmetic hierarchy), i.e., $I\Sigma_n \vdash \exists y \Psi(0, y) \wedge \forall x (\exists y \Psi(x, y) \rightarrow \exists y \Psi(x + 1, y))$ but not $I\Sigma_n \vdash \forall x \exists y \Psi(x, y)$. Define

$$J(x) \Leftrightarrow \exists s (Seq(s) \wedge lh(s) = x \wedge \forall (i < x) \Psi(i, (s)_i)).$$

To verify that $J(x)$ is Σ_{n+1} one need only check that Seq and lh are recursive, Φ is Σ_{n+1}, and the universal quantifier is bounded. Let $(s)_i = y$ such that $I\Sigma_n \vdash \Psi(i, y)$. $I\Sigma_n \vdash J(0)$ and $I\Sigma_n \vdash \forall x (J(x) \rightarrow J(x+1))$ follow from $I\Sigma_n \vdash \exists y \Psi(0, y) \wedge \forall x (\exists y \Psi(x, y) \rightarrow \exists y \Psi(x + 1, y))$. And $I\Sigma_n \vdash \forall x \forall y (y < x \wedge J(x) \rightarrow J(y))$ is immediate: If s is the sequence of length x, let t be the subsequence of s with the same first y elements so that $Seq(t) \wedge lh(t) = y \wedge \forall (i < y) \Psi(i, (t)_i)$. But if $I\Sigma_n \vdash \forall x J(x)$, then $I\Sigma_n \vdash \forall x \exists s \Psi(x, (s)_x)$, a contradiction. Thus J is a proper $I\Sigma_n$-cut. A similar construction works also for $n = 0$, but one needs more creative codings for sequences.

Intuitively one may think of a cut of an arithmetical theory's numbers as an initial segment of a model of that theory. But this intuition is somewhat inaccurate: Not all initial segments of a theory's models are necessarily definable by a predicate, and it is possible for a theory to have a syntactic cut that does not define any tangible geometric structure treatable outside the theory.

Syntactic cuts can be used to interpret relatively strong theories in weaker ones. For instance Q does not prove the commutativity of addition. It may, however, prove the commutativity of addition for all numbers that fall under a predicate J, i.e.,

$$Q \vdash \forall x \forall y (J(x) \land J(y) \rightarrow x + y = y + x). \qquad (5.1)$$

Of course the above formula is trivially provable in Q when J is empty. There are, however, predicates $J(x)$ that contain 0, are closed under successor and addition, and for which the above formula is provable. Since Q does not prove the commutativity of addition for all natural numbers, Q must not prove induction over $J(x)$ if Q is consistent, so such predicates define proper syntactic cuts for Q's numbers. In such a case one says that Q proves the commutativity of addition "with quantifiers restricted to J." If one defines the commutativity of addition by the formula

$$Commute_+ \Leftrightarrow \forall x \forall y (x + y = y + x),$$

then the above formula (5.1) with quantification restricted to J is abbreviated $Commute_+^J$.

It is possible for a theory to prove, not just properties of arithmetical operations that it otherwise cannot prove, but also axioms of stronger arithmetical theories by restricting quantifiers to suitable predicates. For example, Wilkie and Paris [1987] and Nelson [1986] independently showed that there is a cut I for which

$$Q \vdash \Phi^I \quad \text{for all theorems} \quad \Phi \quad \text{of} \quad I\Delta_0.$$

In this case we say that the predicate $I(x)$ is an interpreting domain for $I\Delta_0$ in Q: One can emulate $I\Delta_0$ in Q by quantifying only over numbers in J. This relation is expressed $Q \overset{I}{\succeq} I\Delta_0$. Wilkie and Paris have also shown [1987] that there is a subcut J of I (i.e., $J(x)$ is a cut and $Q \vdash \forall x (J(x) \rightarrow I(x))$) for which

$$Q \vdash \forall x \forall y \exists z (J(x) \land J(y) \rightarrow J(z) \land z = x^{(\log y)}).$$

This last theorem of Q is the totality of the ω_1 function with quantifiers restricted to J. Since $Q \vdash \forall x (J(x) \rightarrow I(x))$, it follows that $Q \vdash \Phi^J$ whenever $Q \vdash \Phi^I$. Hence

$$Q \overset{J}{\succeq} I\Delta_0 + \Omega_1.$$

Pudlák's theorem is that for no cut J does $Q \vdash Con_Q^J$. This means, in particular, that we not only have $Q \nvdash Con_Q$ but also $Q \nvdash Con_Q^J$ for J such that $Q \overset{J}{\succeq} I\Delta_0 + \Omega_1$. That is, Q cannot prove even that there is no Q-proof of contradiction whose Gödel number lies in an interpreting domain for $I\Delta_0 + \Omega_1$.

This strengthens the theorem of Bezboruah and Shepherdson. Extensionally, their result shows that Q cannot rule out the possibility that one of its theorems is $\bot$. Wilkie and Pudlák's results together show that Q cannot rule out $\bot$ even from among just those theorems whose Gödel numbers are in a model of $I\Delta_0 + \Omega_1$. Intuitively Pudlák's argument is that Gödel's theorem in its classical form might be meaningless for Q because such a weak theory does such a poor job of defining the natural numbers that it is no wonder it cannot rule out a proof of $\bot$ – it after all has to consider a rather wild array of nonstandard "pseudonumbers." But if it is consistent with the axioms of Q even for there to be a proof of $\bot$ coded in a model of $I\Delta_0 + \Omega_1$ that Q defines, one might be inclined to understand this as "the unprovability in Q of Q's consistency" because models of this stronger theory are more well behaved.

Exactly how well behaved must models of $I\Delta_0 + \Omega_1$ be for this strengthening to be meaningful? Pudlák says "In such a cut all reasonable definitions of Con_Q are equivalent." In Section 5.3 we questioned whether this is true, i.e., if the idea is that all definitions of "the consistency of Q" are equivalent, then the unprovability in bounded arithmetic of the biconditional $Con_Q \leftrightarrow HCon_Q$

draws this into question. One might ask further, even if this were true, whether it would be relevant to the metatheoretical meaning for Q of Wilkie and Pudlák's theorems. Pudlák's suggestion can be made precise by pointing out that models of $I\Delta_0 + \Omega_1$ are sufficiently well behaved because the theory proves the Feferman criteria for theorem-hood. His implicit argument is prescriptive in this sense: Since the predicate Thm_Q is intensionally adequate in $I\Delta_0 + \Omega_1$, one need only formulate provability with the variant predicate $Thm_Q^J(x) \leftrightarrow J(x) \wedge Thm_Q(x)$ to attain intensional adequacy in Q.

The intensionality of an arithmetization of syntax for Q, even with the variant predicate, is not so immediate, though. To see this it is instructive to review the proof of Gödel's second theorem. With the intensionality criteria for a theory $T \supseteq I\Delta_0 + \Omega_1$ and the existence of a formula ϕ for which $I\Delta_0 + \Omega_1 \vdash \phi \leftrightarrow \neg Thm_T(\overline{\phi})$[5] the theorem for T proceeds in a standard way:

THEOREM 8 (Gödel)   *If $T \supseteq I\Delta_0 + \Omega_1$ is decidable and consistent, then $T \nvdash Con_T$.*

Proof.   From the first incompleteness theorem for T, $T \nvdash \phi$. $I\Delta_0 + \Omega_1 \vdash \neg\phi \rightarrow Thm_T(\overline{\phi})$. $I\Delta_0 + \Omega_1 \vdash Thm_T(\overline{\phi}) \rightarrow Thm_T(\overline{Thm_T(\overline{\phi})})$ by the fifth intensionality criterion. $I\Delta_0 + \Omega_1 \vdash Thm_T(\overline{Thm_T(\overline{\phi})}) \rightarrow Thm_T(\overline{\neg\phi})$ by the first intensionality criterion and choice of ϕ. It follows that $I\Delta_0 + \Omega_1 \vdash \neg\phi \rightarrow Thm_T(\overline{\neg\phi})$. Hence $I\Delta_0 + \Omega_1 \vdash \neg\phi \rightarrow (Thm_T(\overline{\phi}) \wedge Thm_T(\overline{\neg\phi}))$.

Now by the fifth intensionality criterion $I\Delta_0 + \Omega_1 \vdash Thm_T(\overline{\psi})$ for any tautology ψ, and in particular $I\Delta_0 + \Omega_1 \vdash Thm_T(\overline{\phi \rightarrow (\neg\phi \rightarrow \bot)})$. Also by two applications of the second

---

[5]   See Buss [1998c] §§2.1–2.2 for demonstrations.

intensionality criterion $I\Delta_0 + \Omega_1 \vdash \neg\phi \to Thm_T(\overline{\neg\phi \to \bot})$ and $I\Delta_0 + \Omega_1 \vdash \neg\phi \to Thm_T(\bot)$. Hence $I\Delta_0 + \Omega_1 \vdash Con_T \to \phi$. The theorem now follows by the first incompleteness theorem, since $T \supseteq I\Delta_0 + \Omega_1$.

$\dashv$

It is obvious that one cannot readily generalize this proof for arithmetics not extending $I\Delta_0 + \Omega_1$. Even if one had Gödel's first theorem for such a weak theory, one would not be able to infer ϕ from Con_T, since the conditional $Con_T \to \phi$ would not necessarily be a theorem of T. There is a way around this obstacle, though, as Bezboruah and Shepherdson's proof reveals. Another solution, evident from the present book, is to recast provability in $I\Delta_0 + \Omega_1$ as provability in a weaker theory T relativized to a syntactic cut J for which $T \overset{J}{\succeq} I\Delta_0 + \Omega_1$. Still another difficulty remains, however, which is that the intensionality criteria used throughout the proof might not hold for arithmetizations of weak theories, even with the relativized predicate $Thm_T^J(x)$. This is an obstacle to generalizing the proof under consideration to the setting of very weak arithmetics, since these criteria are cited throughout the proof. More importantly (since we know the theorem *can* after all be proven by a different method) this jeopardizes the meaningfulness of the theorem for theories like Q, in so far as the Feferman criteria were supposed to secure the theorem's semantic content.

For example in order to arrive at $I\Delta_0 + \Omega_1 \vdash \neg\phi \to Thm_T(\overline{\neg\phi})$, the above proof relies on $I\Delta_0 + \Omega_1 \vdash Thm_T(\overline{\phi}) \to Thm_T(\overline{Thm_T(\overline{\phi})})$ which is an instance of intensionality criterion five (and is one of the Hilbert–Bernays–Löb derivability conditions). To relativize the proof one would like to show

$$Q \vdash \neg\phi \rightarrow (J(\overline{\neg\phi^J}) \wedge Thm_Q(\overline{\neg\phi^J})),$$

but one cannot so easily establish

$$Q \vdash J(\overline{\phi^J}) \wedge Thm_Q(\overline{\phi^J}) \rightarrow J(\overline{\phi^J}) \wedge Thm_Q(\overline{Thm_Q^J(\overline{\phi^J})}).^6$$

This last sentence is a derivability condition for Thm_Q^J. It does not follow from a straightforward relativization of the proof because substituting a Δ_1^b definition of Q for T does not yield $I\Delta_0 + \Omega_1 \vdash \forall\overline{A}(Thm_Q(\overline{A}) \rightarrow Thm_Q(\overline{Thm_Q(\overline{A})}))$ but only $I\Delta_0 + \Omega_1 \vdash \forall\overline{A}(Thm_Q(\overline{A}) \rightarrow Thm_\delta(\overline{Thm_Q(\overline{A})}))$ where δ is a Δ_1^b definition in $I\Delta_0 + \Omega_1$ of $I\Delta_0 + \Omega_1$. Furthermore, even if one could show $I\Delta_0 + \Omega_1 \vdash \forall\overline{A}(Thm_Q(\overline{A}) \rightarrow Thm_Q(\overline{Thm_Q(\overline{A})}))$, a straightforward relativization of this would not result in the desired formula, for

$$Q \vdash \forall\overline{A}(Thm_Q(\overline{A}) \rightarrow Thm_Q(\overline{Thm_Q(\overline{A})}))^J \qquad (5.2)$$

abbreviates

$$Q \vdash \forall\overline{A}(Thm_Q^J(\overline{A}) \rightarrow Thm_Q^J(\overline{Thm_Q(\overline{A})})), \qquad (5.3)$$

but from this one may not derive

$$Q \vdash \forall\overline{A}(Thm_Q^J(\overline{A}) \rightarrow Thm_Q^J(\overline{Thm_Q^J(\overline{A})}))$$

because the embedded occurrence of "$Thm_Q(\overline{A})$" in (5.2) and (5.3) is a numeral rather than a formula and consequently does not change in the relativization.

---

[6] In the proof above, we are able to conclude $I\Delta_0 + \Omega_1 \vdash Thm_T(\overline{\phi}) \rightarrow Thm_T(\overline{Thm_T(\overline{\phi})})$ because by the fifth intensionality condition $I\Delta_0 + \Omega_1 \vdash Thm_T(\overline{\phi}) \rightarrow Thm_\delta(\overline{Thm_T(\overline{\phi})})$ and also $I\Delta_0 + \Omega_1 \vdash \forall u Fmla(u) \rightarrow (Thm_\delta(u) \rightarrow Thm_T(u))$ since $T \supseteq I\Delta_0 + \Omega_1$.

In fact, Pudlák's [1996] proof of $Q \nvdash Con_Q^J$ for $Q \overset{J}{\succeq} I\Delta_0 + \Omega_1$ is entirely different from the proof of Gödel's second theorem just presented. It relies on bounds on proof length rather than proceeding directly from derivability conditions for the relativized predicate $Thm_Q^J(x)$. Without an appeal to the intensionality of the formula $Thm_Q^J(x)$, however, it is unclear on what grounds one is to find Gödel's second theorem semantically contentful for Robinson's theory. It *is* possible to prove the Feferman criteria for this formula, though as we have seen the interpretability of $I\Delta_0 + \Omega_1$ in Q and the fact that the criteria hold in $I\Delta_0 + \Omega_1$ alone do not suffice. Since it has been suggested (by Pudlák and also Buss [1998c], p. 118) that the intensionality in Q of $Thm_Q^J(x)$ as a representation of theorem-hood is immediate from these two facts, we now show how actually to prove the criteria in Q.

The proof relies on two additional facts about what $I\Delta_0 + \Omega_1$ proves about Q. The first is that $I\Delta_0 + \Omega_1$ proves that all numbers fall, provably in Q, under all syntactic cuts in Q.

LEMMA 9   *For every formula* $I(x)$ *for which* $Q \vdash I(0) \wedge \forall x (I(x) \to I(x+1))$,

$$I\Delta_0 + \Omega_1 \vdash \forall x Thm_Q(\overline{I(x)}).$$

The next fact says that the Σ-completeness of Q is provable in $I\Delta_0 + \Omega_1$.

LEMMA 10   *For all* Σ_1^b *formulas* $\psi(x, y)$,

$$I\Delta_0 + \Omega_1 \vdash \forall x (\exists y \psi(x, y) \to Thm_Q(\overline{\exists y \psi(x, y)})).$$

The proofs of these two facts involve techniques in bounded arithmetic not developed in this book. They are Lemmas 5.21 and 5.24(ii) of Hájek and Pudlák [1993].

We shall prove only the fifth of Feferman's intensionality criteria for $Thm_Q^J(x)$ for two reasons. (i) The first two criteria do in fact follow from the intensionality of the arithmetization of $I\Delta_0 + \Omega_1$ and the interpretation $Q \overset{J}{\succeq} I\Delta_0 + \Omega_1$, and it is evident how to prove the third and fourth criteria from the proof of the fifth. (ii) The fifth criterion entails the derivability conditions that make the above proof of Gödel's second theorem possible.

THEOREM 11 $Q \vdash Thm_Q^J(\phi) \to Thm_Q^J(\overline{Thm_Q^J(\overline{\phi})})$.

Proof.   First note that since Q has an explicit finite axiomatization, the formulas $Fmla_Q(x)$, $Proof_Q(x)$, $Prf_Q(x, y)$, etc., all are Δ_1^b in Q, since the finite axiomatization of Q ensures a polynomial time decision procedure for them. (On the other hand, by Gödel's first theorem $Thm_Q(x)$ is not even decidable.) One cannot say the same about the relativizations $Prf_Q^J(x, y)$, etc. of these predicates because as the cut J is *proper* in Q it is relatively computationally complex. However, by the $I\Delta_0 + \Omega_1$-formalized Σ-completeness of Q,

$$I\Delta_0 + \Omega_1 \vdash Prf_Q(p, \overline{\phi}) \to Thm_Q(\overline{Prf_Q(p, \overline{\phi})}).$$

Now let J be a cut in Q for which $Q \overset{J}{\succeq} I\Delta_0 + \Omega_1$. By the first lemma,

$$I\Delta_0 + \Omega_1 \vdash Thm_Q(\overline{J(p) \wedge J(\overline{\phi})}).$$

From these two formulas it follows that

$$I\Delta_0 + \Omega_1 \vdash Prf_Q(p, \overline{\phi}) \to$$
$$Thm_Q(\overline{J(p) \wedge J(\overline{\phi}) \wedge Prf_Q(p, \overline{\phi})}),$$

which with the $\exists$ rules yields

$$I\Delta_0 + \Omega_1 \vdash \exists u\, Prf_Q(u, \overline{\phi}) \rightarrow$$
$$\overline{Thm_Q(\exists u(J(u) \wedge J(\overline{\phi}) \wedge Prf_Q(u, \overline{\phi})))},$$

or

$$I\Delta_0 + \Omega_1 \vdash Thm_Q(\overline{\phi}) \rightarrow Thm_Q(\overline{Thm_Q^J(\overline{\phi})}).$$

And only now, by choice of J,

$$Q \vdash Thm_Q^J(\overline{\phi}) \rightarrow Thm_Q^J(\overline{Thm_Q^J(\overline{\phi})}).$$

⊣

It follows by the standard argument that $Q \nvdash Con_Q^J$. What can be said, however, of the semantic content of this theorem? The argument above was that the meaning of consistency in $I\Delta_0 + \Omega_1$ is more ambiguous than it first appears since this theory does not prove the equivalence of two reasonable formualtions of consistency. In the present case, where arithmetization is achieved *via* interpretation of a stronger theory, this ambiguity is compounded. The relativized versions of the standard consistency statement and the statement of Herbrand consistency – the versions with quantifiers restricted to an interpretive domain for $I\Delta_0 + \Omega_1$ – provably separate in Q. The *T-separation* of two (perhaps extensionally equivalent) formulas is the T-provability of one and T-unprovability of the other. The following two results relating Q with the unbounded theory $I\Delta_0 + Exp$ show that the formulas Con_Q and $HCon_Q$ are Q-separated when relativized to a sufficiently short cut. The first is a result of Wilkie for which he gave a model-theoretical proof:

THEOREM 12 (Wilkie) *For bounded formulas $\psi(x)$ the following are equivalent.*

1. $Q \preceq Q + \forall x \psi(x)$
2. $I\Delta_0 + Exp \vdash \forall x \psi(x)$

Proof. We need a proof-theoretical version of this theorem since it is more useful for our analysis. Accordingly we will prove actually the equivalence between 2 and the following condition:

3. There is a syntactic cut $I(x)$ in Q such that $Q \vdash \forall x (I(x) \to \psi(x))$.

This condition is equivalent to 1. Moreover, we shall only use the direction $2 \Rightarrow 3$, so we prove only this entailment. For our proof we will need to cite Wilkie's result that if $I\Delta_0 + Exp \vdash \forall x \psi(x)$ then there is a k such that $I\Delta_0 \vdash \forall x (\exists y (y = 2_k^x) \to \psi(x))$. We assume this result, since we cannot replace Wilkie's original proof with one using only techniques from this book. We also need a lemma of Solovay known as the technique for shortening cuts:

LEMMA 13 (Solovay) *Let* $T \supseteq I\Sigma_1$. *For each* $n \in \omega$ *and each* T-*cut* I, *there is a* T-*cut* J_n *such that* $T \vdash \forall x (J_n(x) \to I(2_n^x))$.

1. Given I, there is a T-cut J such that $J \subseteq I$ and J is closed under addition.
   Proof: Define $J(x) \Leftrightarrow I(x) \wedge \forall y (I(y) \to I(x + y))$. It is easy to check that $J \subseteq I$ and J is a T-cut.
   If $x, z \in J$, then for each $y \in I$, $z + y \in I$. Therefore $x + z \in J$. So $x + z + y \in I$ for each $y \in I$. In particular $x + z + 0 \in I$. So $I(x + y) \wedge \forall y (I(y) \to (I(x + z + y)))$, i.e., $J(x + y)$. Therefore, J is closed under addition.
2. For each $n \in \omega$ there is a T-cut J_n such that $T \vdash J_n \subseteq I$ and $T \vdash \forall x (J_n(x) \to I(2_n^x))$.

Proof (by induction on n):

Let $J_0 = J$ from 1. Then $T \vdash \forall x (J_0(x) \rightarrow I(x))$, so $T \vdash \forall x (J_0(x) \rightarrow I(2^x_0))$.

Now suppose J_n is given. From 1 we know that we may assume J_n is closed under addition, so long as we can show J_{n+1} is.

Define I_{n+1} in T by $I_{n+1}(x) \leftrightarrow J_n(x)$. Then $T \vdash \forall x (J_n(2^x) \rightarrow I(2^x_{n+1}))$ by hypothesis. So $T \vdash \forall x (I_{n+1}(x) \rightarrow I(2^x_{n+1}))$.

I_{n+1} is a T-cut:

(a) $T \vdash I_{n+1}(0)$ because $T \vdash J_n(1)$.
(b) $T \vdash \forall x (I_{n+1}(x) \rightarrow I_{n+1}(x+1))$ because if $x \in I_{n+1}$, then $2^x \in J_n$. And J_n is closed under addition, thus $2^x + 2^x \in J_n$. So $2^{x+1} \in J_n$, and $x + 1 \in I_{n+1}$.
(c) $T \vdash \forall x \forall y (I_{n+1}(x) \wedge y < x \rightarrow I_{n+1}(y))$ because $T \vdash \forall x \forall y (J_n(2^x) \wedge y < x \rightarrow J_n(2^y))$.

Now I_{n+1} is not necessarily closed under addition. But we repeat the procedure from 1 and define: $J_{n+1}(x) \leftrightarrow I_{n+1}(x) \wedge \forall y (I_{n+1} \rightarrow I_{n+1}(x + y))$. Since $T \vdash \forall x (J_{n+1}(x) \rightarrow I(2^x_{n+1}))$ and also $T \vdash J_{n+1}(x) \rightarrow I_{n+1}(x)$, one has $T \vdash \forall x (I_{n+1}(x) \rightarrow I(2^x_{n+1}))$. Moreover J_{n+1} is easily seen to be a T-cut. But unlike I_{n+1}, J_{n+1} is provably closed under addition, as needed.

$\dashv$

Now we may prove $2 \Rightarrow 3$ of Theorem 12. Suppose $I\Delta_0 + Exp \vdash \forall x \psi(x)$. Wilkie showed with a model-theoretical argument that there is a k such that $I\Delta_0 \vdash \forall x (\exists y (y = 2^x_k) \rightarrow \psi(x))$ (we use the definition of the graph of exponentiation in $I\Delta_0$). Now let J be a syntactic cut such that $Q \overset{J}{\succeq} I\Delta_0$. Then we have

$$Q \quad \vdash \quad \forall x (J(x) \rightarrow ((\exists y (y = 2_k^x))^J \rightarrow \psi^J))$$
$$\vdash \quad \forall x (J(x) \rightarrow (\exists y (J(y) \wedge y = 2_k^x) \rightarrow \psi^J))$$
$$\vdash \quad \forall x (J(x) \rightarrow (\exists y (J(y) \wedge y = 2_k^x) \rightarrow \psi)).$$

Now we wish to adapt Lemma 13 to shorten J appropriately. However, the lemma holds only for arithmetics containing $I\Sigma_1$. Inspection of the proof reveals that Σ_1-induction is never used, though, since the induction is in the metatheory and never formalized. The restriction to theories extending $I\Sigma_1$ was only to ensure the existence of the exponential function. However, since exponentiation can be defined even in $I\Delta_0$ via its graph, the technique for shortening cuts is applicable for these theories also. Since exponentiation is not provably total in $I\Delta_0$, the proof serves also to guarantee the existence of appropriate numerals for exponential instances: For each $n \in \omega$ and each syntactic cut K of $I\Delta_0$, there is a cut J_n of $I\Delta_0$ such that $I\Delta_0 \vdash \forall x (J_n(x) \rightarrow \exists y (y = 2_n^x \wedge K(y)))$.

Consider the case where K is improper in $I\Delta_0$ (i.e., $I\Delta_0 \vdash \forall x K(x)$) and $n = k$. Then

$$I\Delta_0 \vdash \forall x (J_k(x) \rightarrow \exists y (y = 2_k^x)).$$

Since $Q \overset{J}{\succeq} I\Delta_0$, it follows that $Q \vdash \forall x (J(x) \rightarrow (J_k(x) \rightarrow (\exists y (y = 2_k^x))^J))$, or

$$Q \vdash \forall x (J(x) \wedge J_k(x) \rightarrow \exists y (J(y) \wedge y = 2_k^x)).$$

Define the Q-cut I by $I(x) \leftrightarrow J(x) \wedge J_k(x)$ so that $Q \vdash \forall x (I(x) \rightarrow \exists y (J(y) \wedge y = 2_k^x))$. Since $Q \vdash I(x) \rightarrow J(x)$ and $Q \vdash \forall x (J(x) \rightarrow \exists y (J(y) \wedge y = 2_k^x) \rightarrow \psi)$, it follows that

$$Q \vdash \forall x (I(x) \rightarrow \psi(x)).$$

$\dashv$

THEOREM 14 (Sheperdson) $I\Delta_0 + Exp \vdash HCon_Q$.

Proof. From our axiomatization of Q construct the open theory Q_{open} by removing all the universal quantifiers from Q's axioms, and replacing the axiom (3) for the successor function with an equivalent (open) axiom for a predecessor function. Clearly, $Q_{open} \succeq Q$. Reasoning inside $I\Delta_0 + Exp$, suppose $\neg HCon_{Q_{open}}$. Then there is some tautological disjunction of substitution instances of $\neg Q_{open}$. This is impossible, though, since a truth predicate for open formulas is definable in $I\Delta_0 + Exp$. Thus $I\Delta_0 + Exp \vdash HCon_{Q_{open}}$.

Define in $I\Delta_0 + Exp$ the formula $RCon_T(n)$ saying "contradiction is not provable in T with a proof of cut-rank n" so that $RCon_T(0) \Leftrightarrow CFCon_T$. By the interpretability of Q in Q_{open}, for every n there is an m such that

$$I\Delta_0 + Exp \vdash RCon_{Q_{open}}(m) \rightarrow RCon_Q(n).$$

Then by a partial cut-elimination theorem in $I\Delta_0 + Exp$ (Theorem 5.17(ii) of Hájek and Pudlák [1993]),

$$I\Delta_0 + Exp \vdash CFCon_{Q_{open}} \rightarrow RCon_{Q_{open}}(m).$$

Thus, for all n, $I\Delta_0 + Exp \vdash RCon_Q(n)$. In particular, $I\Delta_0 + Exp \vdash CFCon_Q$. Finally, by the provability in $I\Delta_0 + Exp$ of the equivalence of Herbrand provability and cut-free provability,

$$I\Delta_0 + Exp \vdash HCon_Q.$$

$\dashv$

COROLLARY 15   *There is a cut K, $Q \overset{K}{\succeq} I\Delta_0 + \Omega_1$, such that $Q \vdash HCon_Q^K$ and $Q \nvdash Con_Q^K$.*

Proof. The formula $HPrf_Q(x, y)$ is bounded (it is Δ_1^b-definable in $I\Delta_0 + \Omega_1$), so $HCon_Q$ is a $\forall$-theorem of $I\Delta_0 + Exp$. Therefore by $2 \Rightarrow 3$ of Theorem 12, there is a cut $I(x)$ in Q such that $Q \vdash \forall x(I(x) \rightarrow \neg HPrf(x, \overline{\bigwedge Q \rightarrow \bot}))$. This last formula is just $Q \vdash HCon_Q^I$. Since subcuts preserve interpretability, choose $K \subseteq I \cap J$ for $Q \overset{J}{\succeq} I\Delta_0 + \Omega_1$. One then has $Q \vdash HCon_Q^K$ and $Q \nvdash Con_Q^K$ for $Q \overset{K}{\succeq} I\Delta_0 + \Omega_1$.

$\dashv$

Thus there is a very sharp separation between the formulas Con_Q and $HCon_Q$ when the domain is duly restricted. In particular, the arithmetization of metatheory in Q *via* relativization is less stable than the arithmetization of metatheory in the theories that Q thereby interprets: Whatever intensionality can be recovered depends arbitrarily on the cut one chooses for one's interpretation. If, as Pudlák suggests, restricting of the domain of quantification increases the metatheoretical meaning of formulas,[7] then there arises the paradoxical situation where an increase in semantic content results in a precisification of how nebulous a notion theoretical consistency is. In particular in *Robinson Arithmetic* if one relativizes one's arithmetization so as to maximize interpretive strength, whether or not "consistency" even is

---

[7] In the context of Herbrand provability, one can say concretely why restricting the domain of quantification makes a formula more meaningful (although the idea remains vague in the context of the standard formulation). For when one considers substitutions only of terms built up from functions provably total in T, then "provability" in this restricted sense is closed under cuts. That is, if T proves that there is a restricted Herbrand T-proof of Φ, then given any T-cut J T proves that there is a restricted Herbrand T-proof of Φ already in J (see, for example, Krajíček and Takeuti [1992]). Thus restricting quantification to syntactic cuts approximates the restriction that only provably recursive functions be used in the construction of Herbrand disjunctions. In Chapter 4 I argued that this guarantees intensional correctness, because the "strategies" used to verify or falsify formulas are reduced to those surveyable in T.

provable or unprovable depends on exactly how one phrases the question.

The foregoing was a defense of Kreisel's intuition about the meaninglessness of Gödel's second theorem in the very weak arithmetic Q. The argument was that attempts to "express" metatheoretical claims with only the expressive power of Q are ambiguous and therefore unsuccessful. Even in bounded arithmetic, whether or not an arithmetization of metatheory is intensionally correct depends on the route to arithmetization one chooses, and the interpretability of bounded arithmetic in Q only makes matters worse by generating some consistency statements that are unprovable and others that are provable.

I conclude by pointing out that along the way our analysis has vindicated a more general sentiment of Kreisel concerning the philosophical significance of constructivity:

A constructive theorem is a true quantifier-free formula whose non-logical constants are recursive ... Since the only logical constants are the propositional connectives and the non-logical constants are effective, the truth functional interpretation of the propositional connectives is not problematical. The (or a) constructive content of non-recursive formulae will be expressed by means of constructive ones; the purpose of the so-called finitist or constructive consistency proofs of a system consists, for us, not in the allegedly greater "evidence" or "reliability" of constructive proofs compared with non-constructive ones, but in this: they help us to keep track of the constructive content of the steps in the (non-constructive) proofs of the system considered. ([1958], p. 156)

Kreisel reports that one value of tracking "the constructive content of the steps" in a proof is the sharpening of results and discovery of new theorems. Our analysis has uncovered another. Herbrand's

Fundamental Theorem allows one to express provability in purely constructive terms. As Kreisel notes, appropriate controls on the degree of constructivity can assure one that those terms are manageable in specific formal systems. Thus Herbrand's theorem gives one reason to believe that the Feferman criteria for intensional correctness are unsatisfactory in weak arithmetics, for this theorem and Gödel's second theorem together show that the standard formulation of consistency, though it satisfies the Feferman criteria, is not equivalent to a no-counterexample claim interpretable as such in bounded arithmetic itself. On the other hand, especially in arithmetics as weak as Robinson's, the unprovability of anything like the Feferman criteria for an adequately constructive no-counterexample formulation of provability suggests that these systems cannot speak meaningfully of their own provability at all.

The point stressed here is that these findings do not turn on the especially primitive or epistemologically secure status of any principles of finitism or constructivity and that in order to appreciate them one need not subscribe to "constructivist" theses. Limits on methods are used above only to ensure that no recourse to analysis outside of a theory is needed to verify the meaning of an expression within a theory. Such "intensional" expressions are essential if one is to make sense of the intended meanings of results like Gödel's second incompleteness theorem. This particular result, for example, is supposed to demonstrate that a theory, if consistent, does not prove its own consistency. An essential step in extracting this meaning from any formal theorem involves first verifying that the theory even can speak of its own consistency, and this in turn requires analyzing it from a position where the consistency is not already assumed – on the theory's own terms.

CHAPTER 6

Autonomy in context

Recent philosophical writing about mathematics has largely abandoned the *a priorist* tradition and its accompanying interest in
grounding mathematical activity. The foundational schools of the
early twentiethth century are now treated more like historical attractions than like viable ways to enrich our understanding of mathematics. This shift in attitudes has resulted not so much from a
piecemeal refutation of the various foundational programs, but
from the gradual erosion of interest in laying foundations, from our
culture's disenchantment with the idea that a philosophical grounding might put mathematical activity in plainer view, make more
evident its rationality, or explain its ability to generate a special sort
of knowledge about the world.

Three examples of influential philosophical programs illustrate
this general trend. Penelope Maddy's description of mathematical naturalism has drawn so much attention and approbation not
because it rests on a refutation of "first-philosophy" approaches to
mathematics, but because it quietly leaves those approaches behind.
By instead explaining the philosophical significance of methodological problems that arise internal to mathematical practice and taking
seriously the principles that mathematicians follow to address them,
she has tapped into a concern native to mathematics. Her conviction

that mathematics "should not be subject to criticism from, and does not stand in need of support from, some external, supposedly higher point of view" stems from an assumption that an enriched understanding of mathematical activity will more readily arise from a close look at that activity than from measuring it against some other standard ([1997], p. 184). A second example is Haim Gaifman's case for the ability of probabilistic proofs to provide evidence for mathematical theorems.[1] His suggestion challenges classical philosophical theories about the nature of mathematical knowledge, but mathematicians are listening because they are more concerned with attaining mathematical knowledge than with preserving the philosophers' story about its uniqueness and rationality. Similarly, when George Pólya and Imre Lakatos pioneered the philosophical study of the process of mathematical discovery, instead of trying to judge mathematical activity by measuring it against extra-mathematical norms of rationality, they designed their "mathematical heuristics" to foster understanding of successful mathematical thought on its own terms.[2] Mathematicians are naturally interested in a description of their craft that respects its autonomy. They are able to appreciate its philosophical significance from where they are standing, without first having to adopt a foreign attitude about mathematics that draws its legitimacy into question.[3]

Each entry on the growing list of naturalistic views of mathematics has been thought of as radical in its abandonment of the search

---

[1] See, for example, his [2004]. This line of thought grew out of research by Michael Rabin (e.g., [1976]) and has been taken up by Don Fallis in [1997].

[2] See, especially, Pólya's [1954] and Lakatos's [1976].

[3] These three programs are meant to illustrate the variety of "naturalistic" approaches to the philosophy of mathematics, but they are not at all comprehensive. Even more influential have been Quine and Putnam's naturalized approach to studying mathematical ontology (see Quine [1981a] and Putnam [1971]). The work of Reuben Hersh (in [1997] and, with Philip Davis, [1981]) takes yet another approach. The essays in Aspray and Kitcher [1988] canvass still more.

for foundations and in its deliberate inattention to any ahistorical standards to constrain and characterize mathematical activity. Each has had to fight the battle against foundational epistemology anew, to earn respect from philosophers as a reaction to an orthodoxy. But the orthodoxy has few remaining defenders. The perceived need to acknowledge it before casting it aside is a vestige from the era when philosophy tried to model itself as the science underlying all other sciences. That era has passed. No one today expects a discussion of mathematics to begin with disclaimers about why one is not taking seriously Plato's doctrine of recollection. It is merely a matter of time before one will similarly be able to hold philosophically respectful discussions of mathematical thought without being expected to characterize it, to pronounce on its nature, to explain how it is possible, or at least to say why one is not going to try to do any of these things. Philosophy of mathematics has taken a strong anti-foundationalist turn.

The conclusion I want to draw from the recovery and development of Hilbert's thought in this book is multifaceted: First, the dismissive attitude towards Hilbert's thought that Tarski and Hardy kindled in the early twentieth century is obviously fueled by the naturalistic turn that philosophy of mathematics has taken in recent decades. But that is only because the attribution of foundational aspirations to Hilbert has been so influential. Hilbert's own language often obscured his actual aims: His talk of a "new grounding of mathematics" and "absolute certainty" resonate with the general foundational tenor of his day. Tarski, Hardy, and others were understandably suspicious of such talk, and despite all his efforts to the contrary, Hilbert could not get these native mathematicians to see through his foundational vernacular to his anti-foundational motives. The great irony is that Hilbert's actual philosophical views would be welcome to contemporary writers if only they were known. The fact that recent philosophy of mathematics has largely turned against the foundational programs may even be a result of Hilbert's vision. Hilbert hoped that meta-mathematics would make philosophical grounding look superfluous and irrelevant, that by showing how to formulate questions about mathematics within mathematics he would smother any urge to look elsewhere for their answers, and that mathematicians would be able to continue their craft unencumbered by any perceived need either to obey philosophical constraints or to acknowledge philosophical justification. By making all these things possible, meta-mathematics has carved out a space for naturalistic views of mathematical activity to flourish.

The point of the "intensional" reconstruction of arithmetization in Chapters 3–5 is thus not to be revisionary or intrusive. The implementation of standard meta-mathematical constructions has gone a long way, it seems, to quiet our foundational aspirations and re-orient the philosophy of mathematics. My point is, rather, that this has not happened in one resounding argumentative blow, but in subtler, occasional nudges to our communal concerns. Because meta-mathematics has flourished as it has, the ordinary, retail, scientific view that it invites us to adopt when we ask how mathematical techniques work has simply become more interesting than the extraordinary, context-less view that, for all its sublimity, never managed to take our understanding of mathematics very far. But our shift in interest was *so* gradual that we failed to notice it. We carried out Hilbert's prophecy so unknowingly that we could not credit him for setting the whole process in motion. Had the development of meta-mathematics been a more conscious realization of Hilbert's vision, the naturalistic conceptions of mathematics might have emerged without nearly so much struggle, and the analysis available in the mathematical study of mathematical methods might be deeper. Thus my reconstruction of arithmetization is an attempt to piece together a partial picture of how mathematics' self-accounting might have unfolded if Hilbert's philosophical views had been acknowledged or if Herbrand's inadvertent recovery of them had been able to mature. Intensional investigations of metatheory ought not supplant extensional investigations. They ought to compliment them. They may be the more natural extension of Hilbert's focus on mathematical autonomy and purity, and I hope that my suggestion that they are arouses interest in new approaches to studying Hilbert's program. But, more importantly, their presence alongside standard meta-mathematical constructions should be taken as evidence of an untapped potential to devise a plurality of techniques for projecting questions about mathematics into mathematics itself. I am sure that is the purpose to which Hilbert most hoped his invention would be put.

A fitting way to bring this book to a close is to trust that all these suggestions are now well enough in place and to look ahead. I have not tried to issue a final, conclusive word on the structure and goals of Hilbert's program or the philosophical vision that lies behind it so much as to expose the errors behind traditional depictions of Hilbert's thought and to say a few of the first words in a renewal of interest in Hilbert's program and in the philosophical significance of meta-mathematics. But by uprooting Hilbert from the foundational scene of the 1920s and showing that his engagement with the various attempts to ground mathematical activity was reactionary, rather than participatory, I have left Hilbert's thought floating free. A better image of Hilbert's program than what I have sketched in this book will require relocating Hilbert's thought among views more resonant with his own. One project ahead of interpreters of Hilbert's program is thus to compare his conception of mathematical autonomy with the various naturalistic accounts of mathematical practice that have arisen in the wake of the development of meta-mathematics and the decline in interest in foundationalist epistemology.

To this end, as an example of the kind of investigation that might put Hilbert's vision of mathematical autonomy in context, I turn once again to the writing of Ludwig Wittgenstein. Wittgenstein's critique of second-order discourse is a starting point for many naturalistic views in philosophy. The same critique led Wittgenstein to pronounce on what meta-mathematics could and could not accomplish philosophically. Against these pronouncements, Hilbert's own vision for meta-mathematics comes into sharper focus.

6.2 WITTGENSTEIN'S CRITIQUE OF THE SECOND-ORDER

Section 192 of the *Philosophical Investigations* is a remark that is thematic of a central thread of Wittgenstein's discussion:

One might think: if philosophy speaks of the use of the word "philosophy" there must be a second-order [*zweiter Ordnung*] philosophy. But it is not so: it is, rather, like the case of orthography, which deals with the word "orthography" among others without then being second-order.

This remark is startling, for it arises amid Wittgenstein's extended series of remarks about philosophy and about the *Philosophical Investigations* themselves. The person who might think that there must be a second-order philosophy – the person to whom Wittgenstein issues his rejoinder – is not the naive traditional philosopher, the target of Wittgenstein's ironic caricatures and exposures in the bulk of the book. This remark is meant for the dedicated initiate who has worked through the first 191 sections of the *Investigations* and has all of Wittgenstein's self-conscious and self-referential discussion fresh in mind. Wittgenstein supposes that this discussion might justifiably lead one who is immersed in it to think of it as a sort of second-order discourse, a "meta-philosophy" distinct from "philosophy of the first order." (Indeed, the *Philosophical Investigations* are often read as a piece of meta-philosophy.) But Wittgenstein thinks that taking his discussion in this way is an error. Thus he says so in this remark.

At the outset, it is not clear what error Wittgenstein has in mind when he would disavow us of the notion that his remarks are to be taken as second-order remarks. Since they are *about* philosophy, his remarks are at least in this very mundane sense meta-philosophical. Here the orthography analogy is misleading if pressed too earnestly. When Wittgenstein says, for example, that it is not possible to advance theses in philosophy, that philosophy leaves everything as it is, and that the philosopher's task is to show the fly its way out of the fly-bottle, he is not commenting about the word "philosophy" but rather about the activity that we call by that name. But the orthography analogy is not meant to challenge the idea that his remarks could be about philosophy. It is meant to challenge the idea that they are part of a practice different from the one they are about, "a second-order philosophy" whose task is to evaluate a first-order practice, to put it in its place, to describe and understand its methods and tasks without sharing in them.

This is a radical idea. Sections 103–133 of the *Investigations* are not just descriptions of the tasks and methods of philosophy. They are pronouncements of what philosophy ought to be like, what it can hope to accomplish, and what its limits are. And, crucially, they are not pronouncements that Wittgenstein expects will sit comfortably with professional philosophers. So it can seem like an ironic moment in the *Investigations* when Wittgenstein suggests that all this iconoclasm is the product, not of the alien methods of a second-order philosophy, but of philosophy as it has always been practiced, once it is turned against itself.

The irony is diluted somewhat when one observes a tension in Wittgenstein's attempt, here, to turn philosophy against itself. With one hand he is deconstructing the motifs and self-image of ordinary philosophy by leading his audience slowly through its basic moves in order to expose the superstitions behind them. With the other hand he is trying to salvage ordinary philosophy from the wreckage that he has piled upon it because he needs these scraps to launch his deconstruction. Thus it is not philosophy as it has always been practiced that results in the iconoclastic pronouncements, but instead philosophy pruned by this very process.

It is tempting to ask which step in this process should come first – the philosophical description of philosophy itself or the setting aright of philosophical method. For if one starts out with the former, then the result is indeed the ironic one, and it is hard to understand how Wittgenstein would explain how philosophers have failed to arrive at the same description of their activity as he has. And if one begins with the latter, then it is not clear why one should agree with Wittgenstein about what "right method" is. But Wittgenstein is not troubled by this sort of conundrum because his project is largely indifferent to where one is when one embarks on it. His hope is, rather, that his readers will start wherever they are and that through enough iterations of re-description of their activity and re-marshalling of their armory, eventually they will find their ways out of the trappings of philosophical delusion. "The real discovery," he writes at §133, "is the one that gives philosophy peace, so that it is no longer tormented by questions that bring *itself* in question." (I see the persistence of "the interlocutor" as evidence that this is Wittgenstein's conception of his investigations: Topics are re-described and re-approached repeatedly at the interlocutor's insistence, often without there being any evidence that any ground has been gained for us to trace back through the dialog. Wittgenstein is not being pedantic in assuming that we need this much help to get to some point which comes at the end of such a piece of dialog. The dialog itself, the fact that only the act of winding through a process can properly prune philosophy's view of itself, often *is* the point.)

So Wittgenstein, in denying that his pronouncements "about the use of the word 'philosophy'" make up a sort of second-order discourse, is not saying that the ordinary philosophy that they *are* part of is something with which all philosophers are familiar. These pronouncements are part of what philosophy has become, for Wittgenstein and his audience, in the course of the *Investigations*.[4] But still the questions remain: Why shouldn't there be a second-order philosophy? Why must the pronouncements *about* philosophy be launched *from within* philosophy itself? Wittgenstein has two different ways of answering these questions, each intertwined with the broader picture of his investigations.

The more direct answer is that there just is no compelling reason why the practice of getting a clear picture of "the use of the word 'philosophy'" should differ in its methods from the practice of getting a clear picture of how words like "mean," "know," "understand," and "obey" are used. Since Wittgenstein concedes our ability to engage in the former practice – since his misgivings

---

[4] This is an oddity about philosophy: "It leaves everything as it is," except for philosophy itself. Section 116 exemplifies this attitude: "When philosophers use a word – 'knowledge,' 'being,' 'object,' 'I,' 'proposition,' 'name,' – and try to grasp the *essence* of the thing, one must always ask oneself: is the word ever actually used this way in the language-game which is its natural home? What *we* do is to bring words back from their metaphysical to the everyday use."

about meta-philosophy amount just to the claim that that practice is continuous in its methods with the practice that it is directed at – a second-order philosophy would just be superfluous so long as ordinary methods are up to the task. To the objection that this all rests on the peculiar assumption that philosophical tasks are always about getting a clear picture of the use of language, Wittgenstein would reply that this is not an assumption but rather the view of things that his investigations leave us with. This is the force of the beginning of §107: "The more narrowly we examine actual language, the sharper becomes the conflict between it and our requirement. (For the crystalline purity of logic was, of course, not a *result of investigation*: it was a requirement.)" So much the worse for the crystalline purity of the second-order, Wittgenstein suggests, and for *a priori* requirements in general. Even the definitive philosophical urge of the *Investigations* – to draw into question philosophy itself – is not a requirement for Wittgenstein's project. He never attempts to explain, for example, why it is a worthwhile endeavor. Rather, the history of philosophical thought, the wrong turns it has taken due to historical oddities, and its current strange relationship to everyday thought has left us "tormented" by this urge. Wittgenstein merely recognizes it. His project is to undercut it. To the further objection that there must be some grounds that are not part of philosophy itself from which to settle on that view of philosophy over another, Wittgenstein, in his maddening way, would reply that this is a call for a second-order philosophy, something that he is claiming is superfluous.

The more complicated but more important answer is that the idea of a second-order philosophy is not merely superfluous, but incoherent. According to Wittgenstein, our ability to get a clear picture of a practice or phenomenon can only be informed by our actual standards of what getting clear about something amounts to. These standards are neither fixed nor faultless. They are not fixed because our acts of clarification and our ascriptions of clarity determine what the standards are, and those practices are apt to change. In fact, our very application of these standards, by engendering ever clearer pictures of things, is apt to change those practices. They are not faultless because instead of being derived from an analysis of what it must be to get clear about something, they are determined by our imperfect practices. Indeed an allegation that they are erroneous must draw its sting from them if it is to tie in with any practice of alleging error that we are familiar with. (Wittgenstein's remark in §224 – "If I teach anyone the use of the one word, he learns the use of the other with it" – is applicable here.) And were such an allegation to gain strength, these erroneous standards would still have to be the ones we used in the process of patching them up or revising them until the allegation subsided.

It will not do to draw up new standards, then, to bring to bear on philosophy, simply because these standards will be "drawn up" instead of part of the standards that we actually have in place. Only the latter have any real currency because only they hook up with our practices in a way that makes them meaningful. That is why they are all that is available for ordinary philosophy. For the same reason, they are all that are available for the philosophy of philosophy.

Thus the incoherence that Wittgenstein thinks plagues any second-order philosophy is the same incoherence that he thinks plagues the sort of ordinary philosophy that he criticizes – the "frictionless" sort of endeavor that distorts everything it touches because it does not tie in with our wide range of crude practices, and that does not tie in with those crude practices because it was deliberately set up not to. Wittgenstein puts this traction metaphor to dramatic effect at the end of §107:

We have got on to slippery ice where there is no friction and so in a sense the conditions are ideal, but also, just because of that, we are unable to walk. We want to walk: so we need *friction*. Back to the rough ground!

This diagnosis of the incoherence of second-order philosophy ultimately had little to do with the fact that it was *philosophy* that one got mixed up about when trying to bump it up an order. In quite the same way Wittgenstein tries to dispel the notions that some second-order rule can fix the interpretation of a rule (§§145–152, 185–201), or that a second-order mental image or sign can settle questions about the application of a mental image or the meaning of a sign (§§139–141). The first of these is the notion onto which one is inclined to latch when confronted with the discussion about the free play in how to take a rule. A sketch of Wittgenstein's treatment of it will complement the above discussion of second-order philosophy.

Wittgenstein claims that just as there is nothing "in the rule" determining how it is to be interpreted (cf. §§185–186), there can be no other rule governing how the first is to be interpreted that is not plagued by the same interpretive free play (cf. §§189–190). At first, this does not seem like a new insight, but only a repetition of the original allegation. But in fact an added layer of profundity emerges in Wittgenstein's rehashing of the same point at the second-order. The person grasping for the second-order rule is supposed to have conceded to Wittgenstein the free play in ordinary rules – i.e., rules that share a natural history with our broader set of practices and projects – but to have not yet conceded the same point for rules drawn up specifically to fix the interpretations of other rules.

The hope is that these second-order rules, because they are not tied to a larger body of practices, will be able to have their interpretation built in to them (like the image in §141 of a cube together with its method of projection — the sign that depicts its own meaning). Wittgenstein's point is that either they will be tied in to the same body of practices and for that reason suffer the same fate as ordinary rules despite their being rules about how to interpret rules, or they will not hook up with anything else and for that reason will not have enough "friction" to play the role of rules in our lives. In §192 he calls such frictionless rules "super-expressions" (*Über-Ausdrucken*). Thus by rehashing the point about interpretive free play Wittgenstein takes down a new target, the very notion of a second-order rule.

In the same way Wittgenstein is poised to dispel *any* second-order discourse. Such discourse is presumably meant to accomplish some purpose, and Wittgenstein is prepared to investigate in each case whether the ordinary discourse which the former is about is not already fit to be put to that purpose. For Wittgenstein this question is not to be answered by attempts at a new and ingenious use of that discourse — it is to be read out of a laborious study of the discourse's natural history. Thus there are two possibilities. If the discourse proves up to the task, then it will be our satisfaction with its success that determines that it is, and the task will seem to have been "deflated" compared to the extraordinary vision that one might have formerly cast it under. Otherwise, rather than deeming the ordinary practice insufficient for the task, the task itself will have been exposed as incoherent, as inspired by a distorted and unrealistic view of language.

The philosophy and rule-following discussions exemplify these two alternatives. In the case of philosophy, the tasks are both to describe and, if need be, to amend a practice. If the practice needs amending, then this is because it yields erroneous descriptions and recommends needless emendation, so it appears to be an awful tool to put to this task. But, Wittgenstein claims, it is the only tool we have, and reiteration of the self-description can still be gainful. Our initial response to this claim is supposed to be that the gain is illusory because there are no guarantees that the whole cycle of re-description and emendation will not run off the rails. To quiet this reaction Wittgenstein suggests that the notion of error underlying that worry is itself parasitic on the perceived sublimity of a second-order view of things. He wants us to accept that there are no "rails" other than the ones we lay as we go. In the case of rules, the task is to interpret them, and Wittgenstein is less optimistic that ordinary rules will do the trick. But rather than chalk this up as a victory for second-order rules, he infers that the task is incoherent, that there is *nothing* to point to that fixes the interpretation of a rule. Rule-following takes place within, and therefore cannot be accounted for by, our acts of interpretation.

But the point I want to emphasize is not the ultimate verdict Wittgenstein reaches in any individual case. Indeed, it is not always clear whether the ordinary practice is up to the task or whether the task is illusory because it is often not clear in advance whether the relevant community will accept the task's deflated, naturalistic residue as a sensible replacement for the original version. To illustrate this, whenever Wittgenstein narrows in on the first option, he has his interlocutor accuse him of a naturalistic fallacy. This accusation is then followed by some appeal to the legitimacy of the naturalistic rendering of the task. Wittgenstein always welcomes us to disagree about this last point on condition that we recognize that the inability of ordinary discourse to do what we wanted stems from our having wanted something unattainable, not from any shortcoming of the ordinary discourse. The main point is that it does not matter much to Wittgenstein which of the two conclusions one draws so long as one sees that they are the only two options. Whether ordinary discourse is shown to suffice (as Wittgenstein thinks is the verdict for meta-philosophy) or come up short because of the task's incoherence (as in the case of interpreting rules), the resulting landscape is always purely "first-order." In every case, the second-order discourse is designed to issue pronouncements on an ordinary practice that are in some way purer than what can ever be generated from that practice itself. But in stepping back from the ordinary to judge it, one forfeits the right to judge because one's judgments, however pure, are disconnected from everything that makes language meaningful. "What people accept as justification – is shewn by how they think and live" §352.

To emphasize the systemic presence of this line of thought in the *Investigations*, I conclude by noting that Wittgenstein's critique of second-order discourse is dual to his more familiar attack on "private language." The latter fails to be language because by closing oneself up with one's private experience one has simultaneously cut oneself off from everything that makes language function. One cannot refer to what no one else can experience; one cannot even call it a "something" (cf. §304). Similarly one cannot launch an extraordinary sort of evaluation of the ordinary, because evaluation is and can only be an ordinary practice. Wittgenstein thinks that the notion of mental objects to which private individuals have privileged access and knowledge of which explains our ability to issue incorrigible reports is a vestige of the metaphysical and epistemological myths that are the targets of his explicit criticisms. So too, he thinks, is the notion of an order of discourse whose terms refer to parts of lower-level discourse and whose grammar is specially suited for just this sort of reference.

6.3   FIRST-ORDER META-MATHEMATICS

Hilbert announced his new branch of mathematics at the Hamburg University Mathematics Seminar with these memorable words:

> Just as the physicist investigates his apparatus and the astronomer investigates his location; just as the philosopher practices the critique of reason; so, in my opinion, the mathematician has to secure his theorems by a critique of his proofs, and for this he needs proof theory. ([1922], p. 208)

Later in the same address, Hilbert recast his announcement in a different image:

> in addition to this proper mathematics, there appears a mathematics that is to some extent new, a *metamathematics* which serves to safeguard it by protecting it from the terror of unnecessary prohibitions as well as from the difficulty of paradoxes. In this metamathematics – in contrast to the purely formal modes of inference in mathematics proper – we apply contentual inference; in particular, to the proof of the consistency of the axioms. (ibid., p. 212)

In these two depictions of the meta-mathematical science that Hilbert was forging it is easy to see the seeds of a conceptual tension. If that science is supposed to proceed on the model of the physicist's investigation of his apparatus and of the astronomer's investigation of his own location, then its techniques should not differ in any profound way from the techniques of familiar, everyday mathematics. After all, no more than the orthographer has special techniques to bring to bear on the name of his own profession, neither the physicist nor the astronomer trade in the arcane when they turn their attention to getting their laboratory in order – the point of their housekeeping is to coordinate their instruments with their targets, and they would disrupt this coordination if the standards to which they held their equipment were not continuous with the standards they applied to their experimental data. (What the philosopher does in such self-referential moments, we have seen, is less clear.) But if meta-mathematics is supposed to proceed "in contrast to ... mathematics proper" by applying "contentual" as opposed to "purely formal" modes of inference, then it is sharply discontinuous with everyday mathematics.

It is tempting to dismiss this tension as merely a sign of Hilbert's indecision about what metaphor he preferred. In fact, however, the question about whether meta-mathematical techniques should be continuous with those of everyday mathematics bears directly on the philosophical significance of Hilbert's program. As Wittgenstein puts it in his *Remarks on the Foundations of Mathematics*: "Does it mean passing out of mathematics and returning to it again, or does it mean passing from *one* method of mathematical inference to another?" (IV-4). To answer this question, it is again essential that one get clear about what meta-mathematics is supposed to accomplish. According to Hilbert, its purpose is to "secure" already existing mathematical theorems by warding off methodological prohibitions of the sort that philosophers habitually recommend and the threat of systematic paradox that they cite as grounds for these prohibitions. But oddly, Hilbert cites this purpose to support his conception of proof theory as ordinary mathematics as well as the two-tiered picture. For example, in his Hamburg address Hilbert criticizes Hermann Weyl for creating an "artificially concocted" standpoint from which to evaluate mathematics, emphasizing that the only meaningful evaluation of mathematics will have to be launched from one of the standpoints native to existing mathematics. But later he persists with the image that "the contentual thoughts" that are unique to meta-mathematics "are removed elsewhere – to a higher plane, as it were" so that it is "possible to draw a systematic distinction" between the ordinary techniques of mathematics

proper and the extraordinary ones of meta-mathematics.[5] Since the
tasks of meta-mathematics give rise to *both* images, they apparently do not settle the question in any straightforward way. A
clearer picture of what it would take to accomplish these tasks is
needed.

By now it is pretty clear what the Wittgensteinian response
will be. Wittgenstein would not opt for the two-tiered image of
meta-mathematics and consequently could only envision "securing" the theorems of mathematics in a way that is within the reach
of everyday techniques. But as usual, this move is not very appealing initially. For if the philosophers are waging a prohibitive war
on ordinary mathematical techniques, and the logician spins a story
about how those techniques are secure after all *out of those very techniques*, the philosophers are bound to be pretty unimpressed. Why
should they be compelled to a truce by anything built up from
the same techniques that they claim are faulty? Can such a cha-
rade really be deserving of the name "securing" the theorems of
mathematics?

The appeal, then, of the conception of meta-mathematics as a
second-order activity, is that by working with different techniques
"from a higher plane" one can step outside of such justificatory cir-
cles and hope to establish a truce. Thus one finds the early proof
theorists emphasizing that their techniques are "contentual" because
they are applied to the concrete signs of formalized, ordinary mathematics. Wittgenstein replies that this way out of the circle is a way

---

[5] The very way Hilbert stacks the "planes" is provocative. The more familiar image in modern logic is that meta-mathematics is at a lower level than ordinary mathematics. The
idea is that only thus can it escape skeptical attacks and infuse mathematics with security. Hilbert's reverse image is not a disagreement on this point but an emphasis of another
one: Meta-mathematics is secure, not only because its techniques are logically simpler than
ordinary ones but also because its every "thought" is infused with a "content" that makes
it altogether different from a formal science.

out of mathematics altogether. In his *Remarks on the Foundations of Mathematics* he writes:

(When one masters a technique, one also masters a way of looking at things...) Mathematical propositions seem to treat neither of signs nor of human beings, and therefore they *do* not. ... Once more: we do not look at the mathematical proposition as a proposition dealing with signs, and hence it *is* not that. (III-35)

Meta-mathematics looks like mathematics, indeed like a particularly well-behaved fragment of it. But the infusion of its techniques with a second-order character tears it away from the everyday to the fantastic.

Thus, for Wittgenstein, if there is going to be a meta-mathematics then it will have to be developed out of whatever techniques are already lying around and without gilding them with a special character that sets them apart from "the place where they have their natural home" (*Investigations* §116 paraphrased). Again, the skeptic will react to the results of such a meta-mathematics in frustration: Any attempt of this sort to "secure" mathematics will be hopelessly circular. Curiously, Wittgenstein agrees with the skeptic's point, but not with his frustration. He takes his critique of the second-order as a *reductio ad absurdum* of the whole enterprise of trying to secure mathematics. In his *Remarks*, Wittgenstein backs up this attitude with his therapeutic ruminations on the rhetorical question "What does mathematics need a foundation for?" (V-13) which lead to his much discussed suggestion that the threat of contradiction need not be the source of so much anxiety after all. The attitude itself is evident in his discussion of Russell's logic and his strategy to undermine it:

It is my task, not to attack Russell's logic from within, but from without. ... That is to say: not to attack it mathematically – otherwise I should be doing mathematics – but its position, its office. (ibid., V-16)

No amount of mathematical armory could ever implicate a mathematical system, just as it could never secure it. (Neither, of course, could any second-order discourse accomplish either goal – Wittgenstein's "attack from without" takes the form of a dissolution of the second-order aspirations of Russell's logic, and is not itself a second-order critique of that logic.)

Thus the verdict that Wittgenstein reaches about whether the evaluation of mathematics should be first-order or second-order is that it cannot be done at all. Mathematics cannot offer such an evaluation without assuming a sort of second-order character, which is to say that there cannot be a first-order mathematical study of mathematics. But Wittgenstein sees Hilbert's techniques as perfectly ordinary and therefore in no meaningful way meta-mathematical. "What Hilbert does is mathematics and not metamathematics" he wrote in *Philosophical Grammar*. "It is another calculus just like any other" (§319). In his *Remarks* he put the same point this way: "mathematics as such is always measure, never thing measured" (II-75).

But Wittgenstein concluded that ordinary mathematics is not fit for the task that Hilbert tried to put it to only because he misunderstood Hilbert's strategy. He concluded that there could be no meta-mathematics at all in much the same way that he concluded that there is nothing that fixes the interpretation of a rule: Second-order rules (and mathematics) are useless because they are second-order; first-order rules (and mathematics) are useless because the task one wants to put them to is too detached from ordinary purposes to be approached with ordinary techniques. Thus neither is the meaning of a rule nor is the security of mathematics grounded in anything. This line of thought is available, though, only if one conceives of meta-mathematics as an attempt to secure mathematics directly, by providing epistemological foundations.

Since Wittgenstein understood Hilbert's intentions in this way, it is no wonder that he concluded that meta-mathematics would have to be second-order if it would succeed. Hilbert's own intentions were more subtle, though.

A more helpful way to understand Hilbert's conception of meta-mathematics and of how meta-mathematics is to "secure" mathematics follows the other Wittgensteinian option – the option that Wittgenstein himself prefers in the case of meta-philosophy. On this interpretation, Hilbert did not envision meta-mathematics as a second-order discourse. His talk of moving to a higher plane was methodological, not philosophical. He therefore did not intend to secure mathematics by grounding it, as skeptics had hoped to do. Rather, Hilbert aimed to secure mathematics by "deflating" the problems that its critics posed. Bernays described this deflation as the transformation of the "problems that present themselves in the grounding of mathematics" through their being uprooted from the "epistemologico-philosophical domain" and recast "into the domain of what is properly mathematical" ([1922b], pp. 221–2). The skeptical charge that the attempt to secure mathematics in this way is circular draws its sting from the fact that the skeptic is still clutching at the foundational notion of security, and thus mistakes Hilbert's attempts to prove the consistency of various branches of mathematics as attempts to secure them. But Hilbert expected the deflation of philosophical problems about mathematics to secure mathematics regardless of how the naturalized versions of those problems are solved. He expected this because after questions about how and whether mathematical techniques work are transformed into mathematical questions, the legitimacy of mathematics no longer depends on their answers. Hilbert's hope was that the growth of meta-mathematics would simply replace the whole "foundations" movement with a mathematical project, that a first-order meta-mathematics would establish the security of mathematics, not by answering questions about mathematics' foundations, but by supplanting them with mathematical ones.

Wittgenstein insisted that "[t]he *mathematical* problems of what is called foundations are no more the foundation of mathematics for us than the painted rock is the support of a painted tower" (*Remarks* V-13). He intended this as a criticism of Hilbert's philosophy, but Hilbert wanted the world to come to see things this way. Hilbert expected that mathematicians would come to realize that the insights they gain from pursuing meta-mathematics deepen their understanding of mathematics far more than attempts to ground their activity ever did, and that foundational questions would for that reason no longer seem pertinent to them. These expectations have largely been met. There are still skeptics who are unsatisfied with Hilbert's naturalized version of foundational questions, who complain that meta-mathematics does not address the consistency of mathematics in the philosophically important sense. But their numbers are shrinking, their complaints fall flat in conversation with mathematicians, their familiar, worn-out factions are as incapable as ever of devising convincing arguments, and mathematics is the more secure for all this.

Of course Hilbert's approach was not wholly Wittgensteinian. He did not share Wittgenstein's conviction that meta-mathematics would have to be developed out of whatever mathematical techniques are already lying around. Wittgenstein took a quietist stance about the use of first-order techniques to dispel foundational thinking. He felt that a close look at ordinary practices already in place would have to suffice, that an ingenious extension of those practices would be artificial and for that reason could not put the natural practices in plainer view. Hilbert was not content just to look closely at existing mathematics. Since he was not bothered by skeptical worries about mathematics to begin with, he did not need to examine existing mathematics in order to shake such worries. His own deep familiarity with mathematical problem solving assured him that mathematics did not need grounding in anything else. But while for Wittgenstein this assurance is the ideal end of philosophical therapy, for Hilbert it is the beginning of scientific innovation. He saw how to build on existing mathematical techniques in a way that would make the autonomy of mathematics more perspicuous, by showing that philosophers' worries were just ill-formed questions. Hilbert expected that everyone would come to agree that, by recovering those questions within mathematics, he had made them more precise, more meaningful, and perfectly natural.

6.4 EVIDENCE OF AUTONOMY

Naturalized views of a practice come in different strengths of antifoundational fervor. Their weakest varieties are nothing more than a rejection of the idea that the practice can be grounded in *a priori* principles. *A priori* principles are supposed by their advocates to explain how knowledge of abstract subjects can be gained and how activities completely unaccountable to empirical reality can be rational. One naturalistic reaction to this supposition is to suggest that *a priori* principles are not needed because empirical science can offer perfectly fine explanations. Philip Kitcher [1988] and W. V. O. Quine [1981b] have suggested views of mathematics that fit roughly under this rubric.

Other naturalists reject this reaction as just one more attempt at needless grounding. This stronger type of naturalism is not a call to look to nature whenever something needs explained, but a distrust of the idea that all parts of our culture need justification. It is not "more naturalisitc" in the sense that it burrows even further into natural science than does the weaker outlook, but rather in that it takes certain parts of our culture as autonomous. It leaves them free to flourish according to their own standards and to earn their place in culture according to how well they integrate with our other scientific, artistic, moral, and ritual practices, rather than according to the merits that any one of those practices ascribes to it.

Maddy articulates this stronger anti-foundational sentiment in her description of "mathematical naturalism." This is the view, she says, that "extends the same respect to mathematical practice that the Quinean naturalist extends to scientific practice." She elaborates:

It is, after all, those methods – the actual methods of mathematics – not their Quinean replacements, that have led to the remarkable success of modern mathematics. Where Quine holds that "science is not answerable to any super-scientific tribunal, and not in need of any justification beyond the the hypothetico-deductive method," the mathematical naturalist adds that mathematics is not answerable to any extra-mathematical tribunal and not in need of any justification beyond proof and the axiomatic method. Where Quine takes science to be independent of first philosophy, my naturalist takes mathematics to be independent of both first philosophy and natural science ... in short, from any external standard. ([1997], p. 184)

I hope the evidence presented throughout this book makes it clear that Hilbert shared this same, strongly naturalistic conception of mathematics. However, there is a subtle difference between Maddy and Hilbert on this point. Maddy emphasizes that her "mathematical naturalist began within natural science," that "it was only after noting that mathematics seems to be carried out using methods of its own that she elected to study and evaluate those methods on their own terms." This study still takes place within natural science, though. Maddy's moral is simply that "if we are to adhere to our fundamental naturalistic impulse, the conviction that a successful practice should be understood and evaluated on its own terms, then what we say [when we try to study and evaluate mathematical methods] should neither conflict with nor attempt to justify" the mathematics. Hilbert, by contrast, began native both to mathematics and to the view that mathematics was autonomous. He never needed to be convinced that a "naturalized epistemology" of mathematics is just as intrusive as old-fashioned epistemology. Thus Hilbert never needed to learn a moral about how to continue a scientific study of mathematics while still respecting its autonomy. When he decided to investigate questions about mathematics, his first and only impulse was to devise such an inquiry out of mathematical tools. Maddy's "naturalistic philosophy of mathematics," though cautious not to hold mathematical activity to any empirical standard, "takes place within natural science" (p. 202). Hilbert's philosophy of mathematics, so far as I can tell, does not take place anywhere. It is simply an unargued for conviction that mathematics' view of itself is bound to be more illuminating than any external view of it will be, whatever precautions those views may take.

The argumentative burden on strongly naturalistic conceptions of mathematics is then not to refute traditional foundational programs. Wittgenstein was right to point out that this task is impossible. All one can hope to do is to help others lose interest in those programs.[6] Neither is the burden to show that first-order investigations suffice to provide mathematics' foundations. Again,

---

[6] In §245 of the *Investigations*, Wittgenstein writes: "what a mathematician is inclined to say about the objectivity and reality of mathematical facts, is not a philosophy of mathematics, but something for philosophical treatment." A philosophy of mathematics would be something verifiable or refutable, but counter-argument is the wrong reaction for remarks about the nature of mathematics. In his *Lectures on the Foundations of Mathematics* Wittgenstein puts the same point another way. Responding to Hilbert's claim that mathematicians will never be driven out of Cantor's paradise, he writes: "I would say, 'I wouldn't dream of trying to drive anyone out of this paradise.' I would try to do something quite different: I would try to show you that it is not a paradise – so that you'll leave on your own accord. I would say, 'You're welcome to this; just look around you'"(p. 103). This tactic is strongly counter to Hilbert's convictions, vis-à-vis set theory, but exactly the attitude that Hilbert shares with Wittgenstein regarding foundational epistemology.

Wittgenstein correctly observed that they cannot.[7] The burden is to explain why making do solely with first-order investigations is safe, why the conception of mathematics as a foundation-less enterprise does not jeopardize its rationality. Maddy addresses this concern by explaining why her reasons for advocating "mathematical naturalism" are not also reasons for adopting "astrological naturalism" (p. 203). Again, she speaks from the point of view of natural science. Since Maddy's naturalist begins in natural science, it is mathematics' usefulness for scientific purposes that draws her interest to mathematics in the first place. Observing that mathematicians seem to devise their own standards for practicing and developing their discipline – standards that empirical science neither recommends nor makes very good sense of – she chooses to appreciate and study mathematical activity as she finds it, rather than try to rein it in. But if mathematics were not useful for her scientific pursuits, its methods would not earn her respect. She is not so generous in her appraisal of astrology, because in so far as astrology's methods are foreign to her, they also steer it away from any scientific use (p. 204).

I prefer a more general response to the threat of relativism than Maddy's, though – one more in keeping with Hilbert's mathematical nativity. Concerns about relativism arise from a mistaken idea that human activities earn their rationality by aligning with the timeless constraints that only philosophers have ways of discovering. It is clear that these concerns are unfounded as soon as one observes that mathematics' rationality is already evident in its profound ability to tie in with other projects that command our interest. There is no reason to think that its role in science is the only measure of how well it is doing this. Howard Stein asks us to understand

---

[7] This is the realization that led Wittgenstein to say that Hilbert's work does not deserve the name "meta-mathematics."

Hilbert as saying that mathematics' "*sole* 'formal' or 'legal' responsibility is to be consistent," but he adds that "it has also what one might call a 'moral' or 'aesthetic' responsibility: to be useful, or interesting, or beautiful" ([1988], p. 255). Mathematics can thus earn its legitimacy in any of a number of ways, but it is wrong-headed to draw up a list in advance of how it must evolve in order to remain useful, interesting, or beautiful. Mathematicians have not come to where they are by attending to such a list, but by devising, obeying, and continuously adjusting mathematics' own inner logic. Mathematical autonomy is a natural conclusion to draw when one observes with Stein that mathematics "cannot be constrained" by its moral and aesthetic responsibilities, even as its fulfillment of them is the final arbiter of its rationality, any more than can poetry – "poetry is not produced through censorship," and neither is mathematics (ibid.).

The traditional interpretation of Hilbert's program that I have challenged has Hilbert's philosophical vision somehow riding on the results of his new science. There is much wrong with this description, but also some kernel of truth in it. Of course I maintain that Hilbert was not, as he is commonly described as being, trying to demonstrate that modern mathematics is ultimately grounded in finitary reasoning about concrete signs. Like Wittgenstein, he did not think it worthwhile, or even coherent, to look for mathematics' foundations in anything at all. And it is also a mistake to think that Hilbert's conception of mathematical autonomy depends in any way on the verdicts issued from meta-mathematics or even on his ability to invent a stable meta-mathematical science. The relationship between Hilbert's views about mathematics and his development of meta-mathematical techniques is exactly the reverse: Hilbert was only able to envision a mathematical investigation of questions about mathematics because he was already so firmly committed to his naturalistic views. Nevertheless, in two different ways meta-mathematics *has* vindicated his views.

Most obviously, the very fact that the invention of a formidable branch of mathematics fell directly out of Hilbert's insistence on "a clear demand for the solution of the problem [of the consistency of mathematical systems] in the mathematical sense" when everyone around him was striving to preserve the problem's original, philosophical character suggests that this insistence was worthwhile. This is not a proof that Hilbert's views were correct in any philosophically deep sense. It is just evidence that Hilbert was right to stress that attempts to tie mathematics down to one philosophical theory or another would do more harm than good. Attempts to justify mathematical techniques by grounding them in philosophical principles, to whatever extent they succeed, leave mathematics beholden to an external authority and thereby threaten to cripple its growth. By contrast, Hilbert's refusal to think of mathematics in this way led to the development of more mathematics.

More subtlety, the fact that meta-mathematics has gradually shifted our interests from philosophically foundational programs, in the process welcoming naturalistic sentiments similar to Hilbert's own, supports mathematics' autonomy. Again, this is not to say that Hilbert managed in any way to demonstrate the truth of his views – only that meta-mathematics has served the double purpose of showing why they are valuable views to hold and of convincing us that contrary views are comparably idle. This is not to say that Hilbert's views are vindicated simply because they have been influential. The popularity of an idea is not evidence of its correctness. It is rather the way that Hilbert's ideas have exerted their influence that supports them. Thanks to Hilbert, contemporary writers have good reason not to be interested in grounding mathematics. When Hilbert proposed that the consistency of a branch of mathematics is better seen as a mathematical problem than as a philosophical question, he did not show that philosophers were irrational or wrong in thinking that mathematics needed external justification. He merely invited us to look at things differently. This "deflation" of the foundational program was controversial and unconvincing when Hilbert first described it. Had it remained so, this would be a sign that it failed to capture something crucial about mathematics' consistency. But history has shown the opposite. As meta-mathematical techniques developed over the last century, mathematicians turned increasingly more often to those techniques, and increasingly less often to philosophical theories, when they wanted to understand their craft. Naturalistic views do not earn their worth through argument and demonstration, but by replacing old questions with new, more interesting ones that turn out to be more rewarding to pursue. They do not refute the views they replace. They reshape our culture's outlook until those views are unrecoverable.

One unlikely example of a philosopher who abandoned foundationalism after many years of resisting the draw of Hilbert's program is particularly illustrative of this second sort of triumph. I would like to conclude this book with a brief reflection on his change of heart. The philosopher is Herman Weyl. Weyl is known as Hilbert's apostate student, a champion of his own predicativist foundational views and later of Brouwer's Intuitionism. A notable feature of his involvement in foundational studies is that his exploration of Predicativism in particular actually generated a seemingly viable solution to the paradoxes of set theory. He was able to recover a great deal of classical analysis on predicative foundations, avoiding the use of proof by contradiction, the axiom of choice, and the entire theory of infinite sets.

We have seen how unimpressed Hilbert was with these successes. Even more remarkable is that they proved ultimately unpersuasive to Weyl himself. In an essay called "The structure of mathematics" that he wrote after extensive exposure to the development of modern physics, Weyl looked back on his own "first-philosophy" approach to staking out mathematics' foundations with suspicion:

> How much more convincing and closer to the facts are the heuristic arguments and subsequent systematic constructions in Einstein's general relativity, or the Heisenberg-Schrödinger quantum mechanics. A truly realistic mathematics should be conceived, in line with physics, as a branch of the theoretical construction of the one real world, and should adopt the same sober and cautious attitude toward hypothetic extensions of its foundations as is exhibited by physics. ([1949], p. 235)

Weyl saw his own efforts at providing a foundation for mathematics as "unrealistic" compared to the development of modern physics. Neither relativistic classical mechanics nor quantum mechanics derived its mathematical foundation from any philosophical vantage point or *a priori* constraint on what the physical world must be like. Both were shaped by a continuous confrontation with reality, through conjecture, experiment, and subsequent systematization of the results. Weyl's new vision of mathematics being shaped and revised – in its "foundations" just as in its higher reaches – by the outcome of continued mathematical activity preserves the word "foundation" but not its philosophical meaning. The idea of "ultimate foundations," in the form of primitive mathematical principles immune from revision and attesting to the rationality of mathematical practice, is simply unconvincing compared to the image of mathematics evolving on its own terms, devising new principles in the face of new problems, and in doing so defying all attempts to characterize or justify its development. Weyl expressed this view beautifully in an obituary he wrote for Hilbert:

> The question of the ultimate foundations and the ultimate meaning of mathematics remains open; we do not know in what direction it will find its

final solution or even whether a final objective answer can be expected at all. "Mathematizing" may well be a creative activity of man, like language or music, of primary originality, whose historical decisions defy complete objective rationalization. ([1944], p. 550)

The point is not that Weyl recognized the error of his earlier ways. He never stumbled upon a refutation of the idea that mathematics could be secured by a correct exposition of its ultimate foundations. The point is simply that the naturalistic view that he voiced towards the end of his career was almost impossible to maintain decades earlier. The intellectual climate neither invited nor welcomed such ideas. Despite this, Hilbert managed not only to maintain such a view but to put it to use in his invention of meta-mathematics. He recast questions about mathematics within mathematics, put them on all fours alongside other mathematical problems, drained them of their epistemological character, and left future mathematicians to discover that the product of this "deflation" was more worthy of their attention than traditional philosophical questions. As a result, the era of foundational aspirations wound gradually to its end. It became unnatural to think that an external critique of mathematics could in any way justify it, and increasingly plain that mathematics' ability to "defy rationalization" is not something that needs to be addressed, but something worth defending.

References

Adamowicz, Z. and L. Kolodziejczyk [2004] "Well-behaved principles alternative to bounded induction," *Theoretical Computer Science* 322(1): 5–16.

Aspray, W. and P. Kitcher (eds.) [1988] *History and Philosophy of Modern Mathematics*. Minnesota Studies in the Philosophy of Science XI. Minneapolis. The University of Minnesota Press.

Benacerraf, P. and H. Putnam (eds.) [1983] *Philosophy of Mathematics: selected readings*, 2nd edition. New York. Cambridge University Press.

Bernays, P. [1922a] "Die Bedeutung Hilberts für die Philosophie der Mathematik," *Die Naturwissenschaften* 10: 93–9. Translated by P. Mancosu as "Hilbert's significance for the philosophy of mathematics" in Mancosu [1998b], pp. 189–97.

Bernays, P. [1922b] "Über Hilberts Gedanken zur Grundlagen der Arithmetik," *JDMV* 31: 10–19 (lecture delivered at the Mathematikertagung in Jena, September 1921). Translated by P. Mancosu as "On Hilbert's thoughts concerning the grounding of arithmetic" in Mancosu [1998b], pp. 215–22.

Bernays, P. [1923] "Erwiderung auf die Note von Herrn Aloys Müller: Über Zahlen als Zeichen," *Mathematische Annalen* 90: 159–63. Translated by P. Mancosu as "Reply to the note by Mr. Aloys Müller, 'On numbers as signs'" in Mancosu [1998b], pp. 223–6.

Bernays, P. [1931] "Die Philosophie der Mathematik und die Hilbertsche Beweistheorie," *Blätter für deutsche Philosophie* 4: 326–67. Translated by P. Mancosu as "The philosophy of mathematics and Hilbert's proof theory" in Mancosu [1998b], pp. 234–65.

Bernays, P. [1954] "Über den Zusammenhang des Herbrand'schen Satzes mit den Neueren Ergebnissen von Schütte und Stenius," *Proceedings* of the *International Congress of Mathematicians, Amsterdam, September 2–September 9* 2: 397.

Bezboruah, A. and J. C. Shepherdson [1976] "Gödel's Second Incompleteness theorem for Q," *Journal of Symbolic Logic* 41: 503–12.

Brouwer, L. E. J. [1928] "Intuitionistische Betrachtungen über den Formalismus," *KNAW Proceedings* 31: 374–9. Translated by W. P. van Stigt as "Intuitionist reflections on formalism" in Mancosu [1998b], pp. 40–4.

Burail-Forti, C. [1897] "Una question sui numeri transfiniti," *Rendiconti del Circolo matematico di Palermo* 11: 154–64. Translated by J. van Heijenoort in van Heijenoort [1967], pp. 105–11.

Buss, S. R. [1995] "On Herbrand's Theorem," *Logic and Computational Complexity: Lecture Notes in Computer Science* 960, Springer-Verlag: 195–209.

Buss, S. R. (ed.) [1998a] *Handbook of Proof Theory.* New York. North-Holland. 1998.

Buss, S. R. [1998b] "Introduction to proof theory," in Buss [1998a], pp. 1–78.

Buss, S. R. [1998c] "Proof theory of arithmetic," in Buss [1998a], pp. 79–147.

Cantor, G. [1883] *Grundlagen einer allgemeinen Mannigfaltigkeitslehre. Ein mathematisch-philosophischer Versuch in der Lehre des Unendlichen.* Translated by W. Ewald as *Foundations of a General Theory of Manifolds: a mathematico-philosophical investigation into the theory of the infinite* in Ewald [1996c], pp. 881–920.

Cantor, G. [1899] Letter to Richard Dedekind, August 28. Translated by W. Ewald in Ewald [1996c], pp. 936–7.

Carnap, R. [1931] "Die logizistische Grundlagen der Mathematik," *Erkenntnis* 2(2/3): 91–105. Translated by E. Putnam and G. J. Massey as "The logicist foundations of mathematics" in Benacerraf and Putnam [1983], pp. 41–51.

Carnap, R. [1934] *Logische Syntax der Sprache.* Translated by K. Paul as *The Logical Syntax of Language.* New York. Harcourt, Brace. 1937.

Chevalley, C. and A. Lautman. [1931] "Notice biographique sur Jacques Herbrand," *Annuaire de l'Association amicale de secours des anciens élèves de l'École normale supérieure*: 66–8. Translated by W. Goldfarb as "Biographical note on Jacques Herbrand" in Goldfarb [1971], pp. 21–3.

Chevalley, C. [1934] "Sur la pensée de J. Herbrand," *L'enseignement mathématique* 34: 97–102. Translated by W. Goldfarb as "On Herbrand's thought" in Goldfarb [1971], pp. 25–8.

Chuang Tzu [1964] *Basic Writings.* Anthologized and translated by B. Watson. New York. Columbia University Press.

Davis, P. and R. Hersh [1981] *The Mathematical Experience.* Cambridge. Cambridge University Press.

Dedekind, R. [1888] "Was sind und was sollen die Zahlen?" W. Ewald's revision of W. W. Beman's translation, "The nature and meaning of numbers," in Ewald [1996c], pp. 790–833.

Descartes, R. [1641] *Meditations on First Philosophy.* Translated by D. A. Cress. 3rd edition. Indianapolis. Hackett Publishing Company. 1993.

Detlefsen, M. [1986] *Hilbert's Program: An essay on mathematical instrumentalism.* Dordrecht. Reidel.

Detlefsen, M. [1993] "Hilbert's formalism," *Hilbert, Revue Internationale de Philosophie* 47: 285–304.

Ewald, W. [1996a] Introductory note to Peirce [1881], in Ewald [1996b], pp. 596–8.

Ewald, W. [1996b] *From Kant to Hilbert: A sourcebook in the foundations of mathematics,* vol. 1. New York. Oxford University Press.

Ewald, W. [1996c] *From Kant to Hilbert: A sourcebook in the foundations of mathematics,* vol. 2. New York. Oxford University Press.

Fallis, D. [1997] "The epistemological status of probabilistic proofs," *Journal of Philosophy* 94(4): 165–86.

Feferman, S. [1960] "Arithmetization of metamathematics in a general setting," *Fundamenta Mathematica* 49: 37–92.

Fourier, J. [1878] *Analytical Theory of Heat.* Translation of *Théorie Analytique de la Chaleur* (1822). Dover. 2003.

Frege, G. [1884] *Die Grundlagen der Arithmetic.* Translated by J. L. Austin as *The Foundations of Arithmetic.* Evanston. Northwestern University Press. Cited according to sectioning.

Gaifman, H. [2004] "Reasoning with bounded resources and assigning probability to arithmetical statements," *Synthese* 140: 97–119.

Gödel, K. [1931] "Über formal unentscheidbare Sätze der Principia mathematica und verwandter Systeme I," *Monatshefte für*

Mathematik und Physik 38: 173–98. Translated by S. Bauer-Mengelberg as "On formally undecidable propositions of *Principia Mathematica* and related systems I" in van Heijenoort [1967], pp. 596–616.

Goldfarb, W. (ed.) [1971] *Jacques Herbrand: Logical Writings*. Cambridge. Harvard Univeristy Press.

Hájek, P. and P. Pudlák [1993] *Metamathematics of First-Order Arithmetic*. Perspectives in Mathematical Logic. Santa Clara. Springer-Verlag.

Hardy, G. H. [1929] "Mathematical proof," reprinted in Ewald [1996c], pp. 1244–63.

v. Helmholtz, H. [1887] "Zälen und Messen, erkenntnistheoretisch betrachtet," *Philosophische Aufsäʒte, Eduard Zeller ʒu seinem fünfʒigjährigen Doctorjubiläum gewidmet*: 17–52. Translated by M. F. Lowe as "Numbering and measuring from an epistemological viewpoint" in Ewald [1996c], pp. 728–52.

Herbrand, J. [1930a] *Recherches sur la théorie de la démonstration*. Herbrand's doctoral thesis at the University of Paris. Translated by W. Goldfarb, except pp. 133–88 trans. by B. Dreben and J. van Heijenoort, as "Investigations in proof theory" in Goldfarb [1971], pp. 44–202.

Herbrand, J. [1930b] "Les bases de la logique hilbertienne," *Revue de métaphysique et de morale* 37: 243–55. Translated by W. Goldfarb as "The principles of Hilbert's logic" in Goldfarb [1971], pp. 203–14.

Herbrand, J. [1931a] "Sur le problème fondamental de la logique mathématique," *Sprawoʒdania ʒ posiedʒeń Towarʒystwa Naukowego Warsʒawskiego, WydʒiałIII* 24: 12–56. Translated by W. Goldfarb as "On the fundamental problem of mathematical logic" in Goldfarb [1971], pp. 215–71.

Herbrand, J. [1931b] "Unsigned note on Herbrand [1930a]," *Annales de l'Université de Paris* 6: 186–9. Translated by W. Goldfarb in Goldfarb [1971], pp. 272–6.

Herbrand, J. [1931c] "Notice pour Jacques Hadamard." Translated by W. Goldfarb as "Note for Jacques Hadamard" in Goldfarb [1971], pp. 277–81.

Herbrand, J. [1931d] "Sur la non-contradiction de l'arithmetique," *Journal für die reine und angewandte Mathematik* 166: 1–8. Translated by J. van Heijenoort as "On the consistency of arithmetic" in Goldfarb [1971], pp. 282–98.

Hersh, R. [1997] *What is Mathematics, Really?* New York. Oxford University Press.

Hilbert, D. [1899] *Grundlagen der Geometrie.* Translated as *The Foundations of Geometry*, 3rd edition, by E. J. Townsend. La Salle. Open Court. 1938.

Hilbert, D. [1904] "Über die Grundlagen der Logik und der Arithmetik," *Verhandlungen des Dritten Internationalen Mathematiker-Kongresses*, Leipzig, Teubner: 174–85. Translated by B. Woodward as "On the foundations of logic and arithmetic" in van Heijenoort [1967], pp. 130–8.

Hilbert, D. [1922] "Neubergründung der Mathematik. Erste Mitteilung," *Abhandlugen aus dem Mathimatischen Seminar der Hamburgischen Universität* 1: 157–77. Translated by W. Ewald as "The new grounding of mathematics: first report" in Mancosu [1998b], pp. 198–214.

Hilbert, D. [1923] "Die logischen Grundlagen der Mathematik," *Mathematische Annalen* 88: 151–65. Translated by W. Ewald as "The logical foundations of mathematics" in Ewald [1996c], pp. 1134–48.

Hilbert, D. [1926] "Über das Unendliche," *Mathematische Annalen* 95: 161–90. Translated by E. Putnam and G. J. Massey in Benacerraf and Putnam [1983], pp. 183–201.

Hilbert, D. [1928] "Die Grundlagen der Mathematik." Translated by S. Bauer-Mengelberg and D. Føllesdal as "The foundations of mathematics" in van Heijenoort [1967], pp. 464–79.

Hilbert, D. [1931] "Die Grundlegung der elementaren Zahlentheorie," *Mathematische Annalen* 104: 485–94. Translated by W. Ewald as "The grounding of elementary number theory" in Mancosu [1998b], pp. 266–73.

Hilbert, D. and P. Bernays [1939] *Grundlagen der Mathematik*, vol. 2. 2nd, revised edn. Berlin. Springer-Verlag. 1970.

Kitcher, P. [1988] "Mathematical Naturalism," in Aspray and Kitcher [1988], pp. 293–325.

Klein, F. [1908] *Elementarmathematik vom Höheren Standpunkt aus: Geometrie.* Translated by E. R. Hedrick and C. A. Noble as *Elementary Mathematics from a Higher Viewpoint: Geometry.* New York. Dover. 1939.

Kline, M. [1972] *Mathematical Thought from Ancient to Modern Times.* New York. Oxford University Press.

Krajíček, J. and G. Takeuti [1992] "On induction-free provability," *Annals of Mathematics and Artificial Intelligence* 6: 107–26.

Kreisel, G. [1951] "On the interpretation of non-finitist proofs I," *Journal of Symbolic Logic* 16: 241–67.

Kreisel, G. [1958] "Mathematical significance of consistency proofs," *Journal of Symbolic Logic* 23: 159–82.

Kreisel, G. [1985] "Mathematical logic: tool and object lesson for science," *Synthese* 62(2): 139–51.

Lakatos, I. [1976] *Proofs and Refutations*. New York. Cambridge University Press.

Maddy, P. [1997] *Naturalism in Mathematics*. New York. Oxford University Press.

Maddy, P. [2008] "How applied mathematics became pure," *Review of Symbolic Logic* 1(1): 16–41.

Mancosu, P. [1998a] "Hilbert and Bernays on metamathematics," in Mancosu [1998b], pp. 149–88.

Mancosu, P. (ed.) [1998b] *From Brouwer to Hilbert: The debate on the foundations of mathematics in the 1920s*. New York. Oxford University Press.

Nelson, E. [1986] *Predicative Arithmetic*. Princeton. Princeton University Press.

Parikh, R. [1971] "Existence and feasibility in arithmetic," *Journal of Symbolic Logic* 36: 494–508.

Peirce, B. [1870] *Linear Associative Algebra*. Washington DC (lithograph). Reprinted in part in Ewald [1996b], pp. 584–94.

Peirce, C. S. [1870] "Description of a notation for the logic of relatives, resulting from an amplification of the conceptions of Boole's calculus of logic," *Memoirs of the American Academy of Arts and Sciences* 9, pp. 317–78.

Peirce, C. S. [1876] Notes on B. Peirce's [1870], in Ewald [1996b], pp. 594–6.

Peirce, C. S. [1881] "On the logic of number," *American Journal of Mathematics* 4: 85-95. Reprinted in Ewald [1996b], pp. 599–608.

Peirce, C. S. [1885] "On the algebra of logic," *American Journal of Mathematics* 7: 180–202. Reprinted in Ewald [1996b], pp. 609–32.

Peirce, C. S. [1902] "The simplest mathematics," reprinted in part in Ewald [1996b], pp. 638–48.

Pólya, G. [1954] *Mathematics and Plausible Reasoning*. Princeton. Princeton Univrsity Press.

Pudlák, P. [1985] "Cuts, consistency statements, and interpretations," *Journal of Symbolic Logic* 50: 423–41.

Pudlák, P. [1996] "On the length of proofs of consistency," *Collegium Logicum, Annals of the Kurt-Godel-Society* 2: 65–86.

Pudlák, P. [2004] "Consistency and games: in search of new combinatorial principles," unpublished manuscript.

Putnam, H. [1971] "Philosophy of logic," in Putnam [1975], pp. 323–57.

Putnam, H. [1975] *Mathematics, Matter, and Method: philosophical papers, vol. 1*. New York. Cambridge University Press.

Quine, W. V. O. [1981a] *Theories and Things*. Cambridge. Harvard University Press.

Quine, W. V. O. [1981b] "Things and their place in theories," in Quine [1981a], pp. 1-32.

Raatikainen, P. [2003] "Hilbert's Program Revisited," *Synthese* 137: 157–77.

Rabin, M. O. [1976] "Probabilistic algorithms," in J. F. Traub (ed.) *Algorithms and Complexity: New directions and recent results*. New York. Academic Press.

Ratner, S. [1992] "John Dewey, empiricism, and experimentalism in the recent philosophy of mathematics," *Journal of the History of Ideas* 53(3): 467–79.

Rorty, R. [1976] "Keeping philosophy pure," reprinted in Rorty [1982], pp. 19–36.

Rorty, R. [1979] *Philosophy and the Mirror of Nature*. Princeton. Princeton University Press.

Rorty, R. [1982] *Consequences of Pragmatism*. Minneapolis. The Minnesota University Press.

Rosser, J. B. [1936] "Extensions of some theorems of Gödel and Church," *Journal of Symbolic Logic* 1: 231–5.

Sieg, W. [1985] "Fragments of arithmetic," *Annals of Pure and Applied Logic* 28: 33–71.

Sieg, W. [1986] "Hilbert's program sixty years later," *Journal of Symbolic Logic* 53: 338–48.

Statman, R. [1978] "Bounds for proof search and speed-up in the predicate calculus," *Annals of Mathematical Logic* 15: 225–87.

Stein, H. [1988] "*Logos*, logic, and *logistiké*: Some philosophical remarks on nineteenth-century transformation of mathematics," in Aspray and

Kitcher [1988], pp. 238–59.

Sterne, L. [1759–67] *The Life and Opinions of Tristram Shandy, Gentleman.* Penguin Classics printing of the 1978 Florida Edition. 1997.

Tait, W. [1981] "Finitism," *The Journal of Philosophy* 78: 524–46.

Tarski, A. [1931] "Sur les ensembles définissables de nombres réels, I." Translated as "On definable sets of real numbers" in Tarski [1956], pp. 110–42.

Tarski, A., A. Mostowski, and R. M. Robinson [1953] *Undecidable Theories.* Amsterdam. North-Holland. 1953.

Tarski, A. [1956] *Logic, Semantics, Metamathematics: Papers from 1923 to 1938.* Translated by J. H. Woodger. Oxford. Oxford University Press.

Tarski, A. [1995] "Some current problems in metamathematics," *History and Philosophy of Logic* 7: 143–54.

van Heijenoort, J. (ed.) [1967] *From Frege to Gödel: A sourcebook in mathematical logic 1879-1931.* Cambridge. Harvard University Press.

Weyl, H. [1921] "Über die neue Grundlagenkrise der Mathematik," *Mathematische Zeitschrift* 10: 37–79. Translated by B. Müller as "On the new foundational crisis of mathematics" in Mancosu [1998b], pp. 86–118.

Weyl, H. [1944] "David Hilbert: 1862-1943," *Obituary Notices of Fellows of the Royal Society* 4(13): 547–53.

Weyl, H. [1949] "The structure of mathematics," Appendix A of H. Weyl, *Philosophy of Mathematics and Natural Science.* Princeton. Princeton Univresity Press. 1949.

Whitehead, A. N. [1898] *A Treatise on Universal Algebra, with Applications.* Ithaca. Cambridge University Press.

Wilkie, A. J. and J. B. Paris [1981] "Δ_0 sets and induction," *Proceedings from the Jadswin Logic Conference*, Leeds University Press: 237–48.

Wilkie, A. J. and J. B. Paris [1987] "On the scheme of induction for bounded arithmetic formulas," *Annals of Pure and Applied Logic* 35: 261–302.

Willard, D. [2002] "How to extend the semantic tableaux and cut-free versions of the second incompleteness theorem almost to Robinson's arithmetic Q," *Journal of Symbolic Logic* 67: 465–96.

Wittgenstein, L. [1939] *Lectures on the Foundations of Mathematics.* C. Diamond (ed.). Chicago. The University of Chicago Press. 1976.

Wittgenstein, L. [1953] *Philosophical Investigations.* G. E. M. Anscombe and R. Rhees (eds.). Translated by G. E. M. Anscombe. Malden, Massachusetts. Blackwell. 2nd edition, 1958. Cited according to sectioning.

Wittgenstein, L. [1956] *Remarks on the Foundations of Mathematics*. G. H. von Wright, R. Rhees, and G. E. M. Anscombe (eds.). Translated by G. E. M. Anscombe. Oxford. Blackwell. 3rd edition, 1978. Cited according to sectioning.

Wittgenstein, L. [1969] *On Certainty*. G. E. M. Anscombe and G. H. von Wright (eds.). Translated by G. E. M. Anscombe and G. H. von Wright. New York. Harper and Row. Cited according to sectioning.

Wittgenstein, L. [1974] *Philosophical Grammar*. R. Rhees (ed.). Translated by A. Kenny. Berkeley. University of California Press. Cited according to sectioning.

Zach, R. [1998] "Numbers and functions in Hilbert's finitism," *Taiwanese Journal for Philosophy and History of Science* 10: 3–60.

Zach, R. [2003] "The practice of finitism: epsilon calculus and consistency proofs in Hilbert's Program," *Synthese* 137: 211–59.

Index

Milton Keynes UK
Ingram Content Group UK Ltd.
UKHW041519181024
449640UK00003B/19